Shucks, Shocks, and Hominy Blocks

Shucks, Shocks, and Hominy Blocks

CORN AS A WAY OF LIFE IN PIONEER AMERICA

Nicholas P. Hardeman

Drawings by Linda M. Steele

LOUISIANA STATE UNIVERSITY PRESS
Baton Rouge and London

Designer: Joanna Hill
Typeface: Primer
Typesetter: G & S Typesetters, Inc.
Printer and binder: Thomson-Shore, Inc.

LIBRARY OF CONGRESS CATALOGING IN PUBLICATION DATA

Hardeman, Nicholas Perkins.
  Shucks, shocks, and hominy blocks.

  Bibliography: p.
  Includes index.
    1. Corn—United States—History.   2. Frontier and pioneer
life—United States.   I. Title.
SB191.M2H36   338.1′7315′0973   80-26534
ISBN 0-8071-0793-X

To the memory of my parents, Glen H. and Marion E. Hardeman, who taught me by word and careful example the old ways of corn culture.

Contents

Illustrations

Preface

Perhaps it was because maize, or Indian corn, was the most dominant product of pioneer life that writers of the times neglected to describe its broad influence on the American culture. Why take the trouble to explain what everyone knows? The corn plant's origins are unknown, but the archaeology, the theoretical and known technical aspects of its genetic development, as well as the Indians' ways of growing and using maize have received detailed attention in many publications. Likewise, corn's major role in today's agriculture has been well researched and recorded by a host of writers.

It is the middle or pioneer period that has remained neglected. Yet corn culture was filled with human interest and drama as it developed into an art, a legend, a ritual, and indeed a way of life. The fragments of the story have been badly scattered, and invasions of technology over the last half-century have all but blotted out the record. This study is intended to revive and preserve that forgotten mainstream of corn lore. My intent is to do more than conduct a nostalgic trip into America's past; I have attempted to portray corn in its setting as staff of life and style of life; as culture, art, and craft; as a unifying family purpose, which became a national driving force.

I owe a debt of gratitude to many individuals and a number of institutions for their invaluable help on this volume. Along with her numerous other contributions, my wife, Ada Mae, of the University of California, Irvine, read and reread the manuscript and made many valuable editorial suggestions. My colleague in western American history, Robert W. Frazer of the University of Wichita and California State University, Long Beach, read and commented very helpfully on the entire work, as did the illustrator, Linda Steele, formerly of California State University, Long Beach. Warren Beck, California State University, Fullerton, and Gloria Ricci Lothrop, California State Polytechnic University, Pomona, helped appreciably with their suggested inclusions.

A number of time-tested corn farmers from the Midwest provided important insights into the ways of corn farming, old and new. They include: R. M. Ferguson of Altus, Oklahoma; Wilbur Mason of Santa Ana, California; Marvin and Jane Freitag of Anaheim, California; Glen V. Conner of Auguanga, California; Walker E. and Ruby Hardeman of Pacific, Missouri; Howard D. Hardeman of Cape Girardeau, Missouri; and Robert C. and Townsend H. Fenn of Sunnyvale, California. Others who gave valuable assistance were Marion Johnson, Berta Potter, and Patricia Etter of Long Beach, California, and Louise Rogers of Big Bear City, California.

The California State University and Colleges provided a timely sabbatical leave. Among the many depositories of historical materials, those most extensively used were University of California libraries at the Berkeley, Los Angeles, and Davis campuses; Henry E. Huntington Library, San Marino, California; State Historical Society of Missouri, Columbia; Missouri Historical Society, Saint Louis; Tennessee State Library and Archives, Nashville; Smithsonian Institution, Washington, D.C.; Houston Public Library; State of New Mexico Records Center and Archives, Santa Fe; Texas State Archives, Austin; University of Texas Library and Archives, Austin; California State Library, Sacramento; and particularly the library at California State University, Long Beach. R. G. Walther of the Smithsonian Institution and Donna Longstreet, Leslie Kay Swigart, John Ahouse, and John Dorsey of California State University, Long Beach, were especially helpful. Most important, Linda Steele, Suzanne Forgeron, Juanita Knox, and Barbara Barton in the California State University, Long Beach, Interlibrary Loan Office, shortened my research efforts by hundreds of hours and thousands of miles by their efficient and cheerful tracking down of obscure and valuable volumes from all corners of the nation.

Shucks, Shocks,
and Hominy Blocks

CHAPTER I

The Monarch

But now again my mind turns to the glorious corn. See it—look on its ripening, waving field. See how it wears a crown, prouder than monarch ever was . . . the royal corn within whose yellow heart there is of health and strength for all the nations.

RICHARD J. OGLESBY, governor of Illinois, 1893

Pioneer America was a many-dimensioned jigsaw puzzle, pieced together in an erratic, westward-bulging panorama. It was length and breadth and elevation; and it was measured in time span and variation of character.

New England was tall-masted ships, whirling water wheels, thrifty homesteads, and soils strewn with glacial rocks. It was blazing fall colors and, everywhere, stone fences fringing small fields of maize or Indian corn.

The colonies and states of the middle Atlantic Coast were deep-cut by river mouths and backed by rolling, dark-wooded ranges. They were studded here and there with shops and farmhouses and blanketed with fields of waving wheat and rustling corn.

South along the Coastal Plain there were wide embayments, Grecian-columned plantation houses, rows of slave quarters, and fields of tobacco, rice, indigo, cotton, and Indian corn. The Piedmont and valleys of the Appalachians were cut by broad clearings, which in turn were slashed by deeply eroded gullies. Timbered ridges stood as silent sentinels over smoke-spewing chimneys and geometrical rows of cornstalks and shocks.

Across the great mountain barrier, the heartland of America was an awesome forest, quilted by oak openings and prairie patches in the Old Northwest and drained by a giant river system with thousands of tributaries. Down the currents drifted an ever-growing armada of flat-

boats carrying tobacco, wheat, and cotton to New Orleans and transporting corn products in the forms of meat, butter, whiskey, cornmeal,
and whole-grain corn.

The western prairies and Great Plains, too, were a cut of pioneer
America, sparsely silhouetted by sod houses, grain elevators, and grazing livestock. All about were amber wheat fields and aisles of corn.
And through much of the southwestern frontier, buttes and rimrocks
looked down, as they had for centuries, upon jagged outcroppings and
reddish soils contrasting sharply with scattered patches of green corn.

"We must not neglect our corn on penalty of starving," wrote an Illinois farmer in 1818. Prominent English author and agricultural scientist William Cobbett (sometimes known as Peter Porcupine), who
farmed in New York State about the same time, paid tribute to corn in
a national context. Crediting "much of the ease and happiness of the
people of the United States" to the "absence of grinding taxation," he
went on to state: "That absence alone, without the cultivation of Indian Corn, would not, in the space of only about a hundred fifty years,
have created a powerful nation." Not only was this plant "the greatest
blessing of the country," according to Cobbett. It was "the greatest
blessing God ever gave to man." Thomas Jefferson, one of the best
known farmers of early America, underscored the critical importance
of the corn crop to his entire plantation early in 1776 after a bad growing season. "My situation on that subject is threatening beyond anything I ever experienced. We shall starve literally if I cannot buy 200
barrels." Most Americans did not have the literacy or the time to record such views for history. They were too busy raising corn.

The measuring scale of history is usually marked by records of great
leaders and important events—political figures, queens, kings, business tycoons, generals and admirals, reformers, wars, bold declara-
tions, pioneering discoveries, revolutions, constitutions, and classic
debates. Too often the less obvious but no less important shapers of the
past are neglected. Indian corn or maize, known simply as "corn" for'
the last century and a quarter, was such a shaping force, a quiet monarch that has reigned supreme in America for thousands of years.
Through the long Indian era before European explorers and colonists
first arrived, through colonial times and revolution, constitutional debates, and the emergence of presidential "giants," it was the silent
master of the predominantly rural scene. From the War of 1812 to
Jacksonian Democracy and Manifest Destiny, through the eras of the
underground railroad and bloody Civil War, corn was as unrelenting

as "Old Man River," as convincing as warriors of an occupational army in dominating the land.

> They are marshalled on the hilltop,
> They are ranked along the plain,
> Their files flash in the sunlight
> Nor heed the cloud nor rain;
> No hush of evening breaks their lines,
> No dawn of dewey morn,
> Vast armies of the prairie,
> The fields of gleaming corn.
> (Mrs. Lyon, "Warriors of the Prairie," 1858.)

This plant was more than a monarch and an occupying force in early America; it was indeed a way of life, and those who would understand and appreciate the meaning of the times must know something of corn and its influence.

Indian corn has established a book of records over the centuries. What more illustrious start than saving both Jamestown and Plymouth colonies from starvation? Either directly or through its role as the chief food for livestock and poultry, it is by a wide margin the largest source of human food in the United States and has been since early colonial times. Nor should its importance as the wellspring for drinks be minimized. Historically it has a long lead over all other crops in the number and significance of its nonfood uses. No other element has caused such a change in the face of the pioneer American landscape, and none other demanded such a high percentage of the country's work time prior to the twentieth century. Not to mention the fact that corn weighed heavily as a social and recreational factor.

Compared to corn, cash crops such as cotton, wheat, tobacco, sugar, rice, and indigo have attracted a disproportionate amount of attention from historians and economists. Yet for 240 years after the settlement at Jamestown, nearly 90 percent of Americans depended mainly on farming for their livelihood, and probably more than 90 percent of these depended heavily upon corn for their survival. Word trickling back to Europe of bountiful American corn crops may have been the greatest single factor in luring new settlers to the colonies and, later, the United States. The only major crop grown in every colony and state, corn was the great common denominator of agriculture. It was too widely raised in America and too little imported by Europe to be a major cash item, but it has always been the country's most important crop. (It was reported that in 1979, soybeans finally passed corn in acreage and cash value.) During the 1850s, the production of wheat, cotton, tobacco, and rice combined had only two-thirds the value of the corn harvest. The census of 1850 showed corn overshadowing wheat six to one in bulk and three to one in value. Even in the South, where cotton was heralded as "king," corn was the staple. As a Mississippian wrote from the heart of the cotton belt in 1854, "most of our farmers rely upon corn, almost solely, for the sustenance of both man and beast." In 1849, the South had 18 million acres planted to corn and 5 million in cotton. These census statistics make a more convincing case for the dominance of corn when it is noted that cotton had been making large gains since the invention of the cotton gin in 1793. Today corn production in the United States is equal to wheat in acreage but more than three times as great in bushels (7 billion of corn compared to 2 billion of wheat annually). Although people eat proportionately less corn directly today than during earlier times, much is consumed indirectly as beef, pork, poultry, eggs, milk, oil, and other products.

During the 5,000 to 6,000 years of its rule over the American crop scene, corn has been drawn into a life-or-death pact with humankind—an alliance which neither can break. Without this golden grain and its many products, the survival of millions of people and the abundant life that other millions now enjoy would be impossible to sustain. Oddly, the reverse is also true. In the domestication process, the secrets of which are locked away in the silence of centuries, corn lost the capacity to reseed itself. Were it not planted, tended, harvested, and stored by human care, it would become an extinct species in a matter of several years. Its tough husk and tight-packed seeds prevent scattering. Therefore, any seedlings that germinate choke each other. Corn and people are bonded together until parted by death.

The key role of maize appeared in many phases of pioneer life. That cabin and clearing of the settler's dream was in reality a cabin and corn crop. Occupying of the frontier was encouraged by laws dealing with "Corn Patch and Cabin Rights," and a claim was usually honored if a corn crop had been planted. The oft-cited land hunger of pioneers was usually corn-land hunger. Indian corn was the first crop planted and the first in production statistics on nearly every family plot. From the initial English settlements onward in time and area to the tilling of the High Plains, survival hinged more on success or failure of corn plantings than on any other product.

Farmer preoccupation with this life-sustaining miracle of the plant kingdom surfaced in the very language of rural America. "How's yer corn doin'?" were the first words spoken after "Howdy, neighbor" when the crusty, isolated sod turners got together, no matter what the reason for the gathering—funeral, wedding, camp revival meeting, or community chore. A pessimistic nineteenth-century farm owner from New Burlington, Ohio, kept a diary for thirty years, mostly about bad weather, frightening plots to undermine the republic, and the sad state of corn crops. As will be seen in Chapter XVII, corn found its way into far more American expressions than did any other crop, and perhaps more than all others combined.

Indian corn spawned its own culture in artifacts as well as language. Those simple, peculiarly American tools and utensils fashioned for coping with corn were more than highly efficient designs for work. Quite unconsciously they were graceful works of art as surely as were the neat checkerboarded rows of stalks and sturdy, waist-banded shocks in the fields.

Late-twentieth-century Americans, when they are reminded of Indian corn, may conjure up fleeting mind pictures of shocks and golden pumpkins under fading sun and harvest moon, or steaming hot corn-on-the-cob dripping with butter, or perhaps girl and boy celebrating the finding of a red-grained ear with blush-faced kisses at a husking bee, or a countrified joke or story. The reality of corn culture is now highly impersonal "factories in the fields," where tanklike tractors, gang plows, and combine harvesters cruise like armored divisions. Only yesterday cornfields were units of semiprimitive production carried on by the hand labor of toiling but close-knit families who, sometimes without quite sensing it, found self-reliance, toughness, compassion for other human beings, and fun in their toil.

Small patches of corn, grown and gathered in the old family style, still huddle in a few isolated corners of the country. But as a culture, a way of life, they are going and almost gone. The old corn culture is a comprehensive history, an index to an all-but-forgotten era. It slipped away so quietly that it has been denied a decent memorial. The dream and romance of an era are not fully sensed and appreciated until they have faded into the past. A people can never fully recapture and return to their former cultures, but if they will, they can hold to their heritage. The heritage of Indian, European, and African America is inseparable from maize or Indian corn.

CHAPTER II

This Fabled Plant

Corn . . . was grown by most pioneer farmers for family use and as a dependable source of feed for livestock. Unlike wheat, it was not attacked by destructive insects and diseases. It could be grown on new ground with a minimum of preparation. It required less labor in harvesting and could be harvested over a longer period. In the absence of hay the stalks and blades could be used as winter fodder. It was indispensable in pioneer farming.

STEVENSON W. FLETCHER,
*Pennsylvania Agriculture
and Country Life,* 1950

What is this marvel of Indian ingenuity, this indispensable grass whose seeds will sprout and grow when over a decade old and will retain food value during more than a thousand years of storage, this bamboolike shoot that acre for acre outproduces wheat nearly three to one on less than one-third the quantity of seed required for wheat? *Maize,* "that which sustains life," it was called by many Indians. Natives of the Virginia-Carolina coast called it *pegatowr.* Other tribal peoples had other names, such as *ewachimneash* (New England Indians), and *ohnasta* (Iroquois). The French labeled it *ble d'Inde.* English settlers in America called it *Indian corn,* and it was widely known by this name until after the middle of the nineteenth century. Some newcomers applied the term *squaw corn* since the grain was raised and processed mainly by Indian women. Now it is just plain *corn* in the United States, although Latin Americans and Europeans frequently refer to it as maize. Swedish botanist Karl Linnaeus supplied its botanical name, *Zea mays,* which is somewhat redundant since the first word came from the Greek *zeo,* "to live."

The plant seemed destined for errors of terminology. *Indian corn* was doubly inaccurate—*Indian* for Columbus' mistaken identification of the continent and its people and *corn* for a broad category of Old World cereal grains which are not at all like the American crop. That well-known Biblical figure of speech, "First the blade, then the ear, then the full-grown corn," referred to wheat or barley. But New World corn has accepted its given name with such vigor that "corn" refers to its sturdy stalks and heavy ears of grain over much of the world.

For all their supposed superstitious beliefs and primitive ways, the American Indians have left modern science baffled by their achievements in domesticating, breeding, and adapting corn. Some botanists have long believed that the plant was developed from a grass called teosinte, perhaps in or near the Tehuacan Valley of Mexico more than five thousand years ago. However, today's specialists have been unable to duplicate the feat. And the Indians who wrought the change with their patience of centuries neglected to pass down the secret if they ever consciously knew it. Perhaps, teosinte is not the parent plant but a hybrid offshoot of a wild corn and some other grass such as *Tripsacum*. Modern experimenters have been able to nearly duplicate teosinte by crossing maize and *Tripsacum*. All three of these plants appear to be related to unknown common ancestors from which they developed by divergent evolution. Discoveries near Guyaquil, Ecuador, in 1976 have cast some doubt on the Mexican-origins theory, indicating that corn may have been developed in South America, and perhaps earlier than the five-thousand-year estimate. Since that discovery a new botanical find has added knowledge, if not new understanding, to the enigma of corn's origin. Rafael Guzmán discovered a hitherto unknown variety of teosinte grass in the Mexican state of Jalisco. Unlike the annual corn plant, it is a perennial, but it has the same number of chromosomes as corn. The current crossbreeding experiments may result in the development of perennial corn, which could be produced with much lower tilling costs. They may also introduce much-needed genetic diversity to corn culture. Botanists are far from agreement as to whether this new-found "close relative . . . is the parent, brother or nephew of corn." That riddle may never be solved, but the most likely theory is that maize was domesticated and broken to hearth from a wild extinct or undiscovered species of a corn or cornlike plant. Recent paleobotanical finds of what appear to be true wild-corn pollens (in building foundation cores from drillings more than two hundred feet deep in Mexico City) may provide the most plausible answer. Paul Mangelsdorf and other scientists have judged these pollens to be of an extinct wild corn dating back twenty-five to eighty thousand years. For well over two centuries, scientists have concluded that there is no living wild corn.

Whatever its ancestors, this overgrown grass is an evolutionist's dream of adaptation. Bunched like a small shrub, it clings to life in semiarid deserts and plains on eight or ten inches of annual rainfall. It grows to a towering twenty feet in steaming, equatorial climates where four hundred inches of precipitation may fall in a year. Corn thrives from Canada and the Dakotas to central Argentina and Chile, from Maine and Florida to Arizona and California, adjusting to altitudes from sea level to terraced plots thirteen thousand feet high in the Andean range. With careful seed selection it adapts very quickly to wide changes in growing seasons and conditions. This hardy stalk pushes its roots outward and downward into the soil sometimes as much as six or seven feet from the base. It has no tap root, but it can draw nutrients from a large volume of earth and shield itself against dry times when it must pump water to quench the thirst of its juicy stalk and milky pearls of grain. When green, corn is about 90 percent water by weight.

Section by section, leaf by leaf, the corn stalk grows from its first tiny, curl-bladed sprout. Depending upon the variety of plant and the environmental conditions, it may extend as few as eight or as many as forty-six blades before reaching maturity. After growing nearly to its full height and starting to unfurl a flowery tassel atop the shiny, sword-like blades, the plant seems to lose its sense of direction, putting forth roots several inches above ground in a hula-skirt shape. These are not merely roots. The stalk's marvelous genetic sense has told it that it will soon be more vulnerable to blasting gales. It must bear the burden of two or more unbalanced ears of grain weighing a pound or more each. These bare roots are braces like flying buttresses on a Gothic tower, which give the corn plant triangulated structural strength for the stresses ahead. Some penetrate the soil and seem to serve as both feeders and braces.

Each corn plant has the capacity to fertilize itself and any other plant of the species. The male blossom, or tassel, extends small branches out in every direction at the top of the stalk so that it can shower pollen on all parts of the plant below. The ear, or female blossom, pushes upward and outward between a blade and the stalk, usually several feet below the tassel. When the ear is very small, long delicate silks appear at its outer end. These filaments, one from each grain, catch pollen from the

Stages of corn growth.

same plant and neighboring ones, and fertilization takes place. Each stalk usually bears two ears, although the number may vary from one to six or seven. A properly fertilized ear may contain five hundred to a thousand grains.

As these seeds mature, the ear grows larger and longer. Its weight bends it downward, conveniently, in time for the drying grains to be protected from rain by a shingle-layered husk, or shuck. This sheath keeps the kernels dry through the heaviest downpours, enabling the farmer to postpone the harvest for many months if necessary, while turning to more urgent work.

The tight husks around the closely packed seeds, above all else, establish the uniqueness of corn in the plant kingdom. Yet the very protection of the hoard of grain dooms it to dependence on people, since, as seen earlier, under natural germinating conditions the hundreds of sprouts would choke each other before any could reproduce.

Although modern science has greatly increased the size of corn ears and the yield per land unit, the kinds and colors of maize are the same as those separated out by Indians countless centuries ago. Scholars' earliest archaeological finds, such as the 5,600-year-old remains at Bat Cave, New Mexico (1948), were of popcorn and pod corn. From these, whose origins are unknown, the Indians developed varieties with large ears and a number of colors. The newer kinds included flint, dent, flour, and sweet corn.

Those ancient ears of pod and popcorn found at Bat Cave were only three-quarters of an inch to one inch long. Pod corn, having a pod or husk around each grain, has assumed little modern significance other than as a biological curiosity. The modern grains, however, still have rudimentary film structures of the earlier pods, branlike particles that can be seen on corn cobs.

Popcorn, on the contrary, has remained important for over five thousand years. Its kernels explode from expansion of their granular structure when suddenly dried by heat. Today, millions of Americans who sit munching popcorn before flickering movie screens, television sets, and fireplaces are following an ancient tradition. Throughout much of the hemisphere generations of Indians popped corn in earthen vessels and ate it around open fires. One-thousand-year-old specimens of the grain from ancient, musty Peruvian tombs still popped when heated! Cortés found the Aztecs of Mexico using popcorn necklaces for themselves and their goddesses and gods.

By a selective process of hybridizing or perhaps by repeated choice of large grains for seed, generation after generation, the kernels of some strains lost their capacity to pop. These grains were still valuable because they could be chewed when parched (cooked and dried). Thus did the first Americans probably develop Indian corn. Possibly the hybridizing or cross-fertilizing resulted from carrying grains on hunting trips, migrations, or forays of war parties. However it was done, the effects were far-reaching.

Through some such evolution, Indians developed the four additional varieties of corn. Flints have smooth rounded grains with a high percentage of hard, flinty starch. Dent corns, called gourd seed by the settlers, have thinner flint casings at the outer ends of kernels. Each grain contracts somewhat and develops a dent in the end when dry-ripe. A mutation that retarded the change from corn sugar to starch was no doubt the beginning of sweet corn, also an Indian contribution. Finally, flour corn, characterized by a high content of soft floury

FLINT   DENT

Two types of corn kernels.

substance at the expense of hard flint, rounded out the half-dozen major biological varieties of maize.

The colors of Indian corn match the principal varieties in number. Here it is very clear that the native peoples (primarily women, who did most of the farm work) deliberately selected, segregated, and bred for predetermined color effects *after* they knew of the color possibilities. Just as leaders of some tribes puffed on pipes and blew ceremonial smoke or dusted corn pollen toward the sun, the four prime directions, and the earth, so Indian maids and maidens often raised different colored corn selectively to meet spiritual needs. They grew red, blue, black, yellow, white, and variegated ears to represent the six cardinal points of direction, north, east, south, west, zenith, and nadir. Religious symbolisms probably varied from tribe to tribe, but the practice reveals that there was some deliberate genetic selection in pre-Columbian America. Of the six colors inherited from the Indians, the pioneers developed more than twenty classifiable shades, varying from purple to tawny, amber, and dark brown, and from copper-red to orange. They soon concentrated on two, however—yellow for livestock and white to grind into meal for human consumption.

More important than the color phases were the countless varieties of corn that settlers developed from the priceless gift of the six basic kinds inherited from the Indians. E. Lewis Sturtevant wrote in the post–Civil War era, "The varieties of maize are numberless, and we know of no adequate attempt to reduce them to a system whereby they may be intelligently classified and described." A mere listing of known varieties would run to pages. The names which echoed around the clearings and cabins from the tongues of early farmers who swore by their favorite "yieldy" types included: Yellow Flint or Golden Sioux, New England Eight-Rowed, King Philip, Browne, Southern Big Yellow, Canada Yellow, Dutton, Yellow Gourd-Seed, Hill, Boone County White, Tuscarora, Narraganset, White Gourd-Seed, Southern Big

White, Northern White Flint, Rhode Island Premium, Long Island White, Mandan, Pueblo, Squaw, Wyandotte, Bloody Butcher, Tom Thumb, White Horse Tooth, Cherokee, Cumberland, Stowell's Evergreen, Darling's Early, Gordon Hopkins Red, Large White Flint, and above all, Reid's Yellow Dent.

Despite the interesting maize-breeding contributions of a number of "gentlemen farmers," two nineteenth-century figures were credited above all others with developing high-yield modern corn. They were John Lorain and Robert Reid. Lorain of Phillipsburg, Pennsylvania, began cross-pollination experiments in the early 1800s. By crossbreeding Big Yellow and White (flint), Little Yellow and White (flint), and Gourd Seed (dent), he developed not only the methods of breeding, but also a corn much higher in yield than flint and more durable than the dent or gourd seed. Although the crossbreeding stages since that time are impossible to trace accurately, it is clear that open-pollinated varieties of the golden yellow corn developed by John Lorain had a dominant influence on corn culture until the introduction of hybrid corn in the 1930s.

An unplanned crossbreeding in the West carried Lorain's contribution much further. The Robert Reid family moved to a new farm in Illinois from Ohio in 1846. They planted a reddish-golden corn known as Gordon Hopkins Red. Because it had not yielded as well for them as they had hoped during the previous season, the Reid family used seed from the Little Yellow variety to plant the "missing hills" where Gordon Hopkins Red did not germinate. Apparently these two corns had an ideal blend of genes that were waiting to be brought together, for the issue from this happy union was Reid's Yellow Dent, the most significant stride in corn production since prehistoric times. Because of its fine showing at the great Chicago Fair in 1893, this was sometimes called World's Fair Corn. Another name commonly applied to this variety, as well as to its several hybrid successors, was Corn Belt Dent. Reid's strain of golden maize became standard in the corn belt of the north-central United States and sharply increased American production. Robert Reid was given the credit, but it was a family achievement, and quite accidental.

There were other corn-breeding milestones, but the greatest single improvement in production was yet to come. By selectively pulling tassels and carefully controlling double-crossed pollination of seed corn, experimenters applied a principle known as "first generation hybrid vigor." The process had been in the developmental stages since experiments by George Shull as early as 1905. In 1933, the year of another great Chicago fair and the low point of the Great Depression, large scale commercial quantities of hybrid seed corn became available, and in less than two decades three-quarters of the corn acreage in the United States was hybrid. The yield over Reid's Yellow Dent was approximately doubled. Today prime corn states such as Iowa produce almost nothing but hybrid corn.

New nitrogen and other fertilizers and a recently developed high-protein strain of corn show promise of greater strides in future production of the golden grain. What was long the most important American input to the food resources of the world has been richly increased. Still, the biological components of which these modern advances have been built were already mature and complete when the Europeans came to America in 1492.

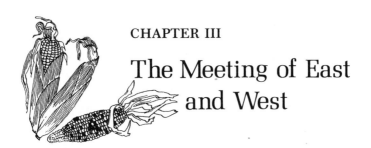

CHAPTER III

The Meeting of East and West

By the old European calendar system it was November 4, 1492. For several weeks an Italian navigator sailing under the authority of Isabella, Queen of Castile, had been prospecting for Asiatic riches in an unlikely sea half a world away from Cathay, or China. Luís de Torres and Rodrigo de Xeres, bearing messages from Christopher Columbus to the Great Kahn, pushed inland on an isle called Cuba. There they came upon two New World crops on the same day that were to cause economic revolutions in many areas of the planet: Indian corn and tobacco.

The hemispheres had met. That frail oceanic bridge of three tiny ships, *Pinta, Nina,* and *Santa Maria,* would be supported and steadily enlarged. Countless other transatlantic gifts would be proffered, among them hardware, glassware, clothing, and smallpox from Europe and beans, squash, potatoes, syphilis, and yellow fever from America. Neither hemisphere would ever be the same again.

Columbus may have taken corn back to Spain on his return trip in 1493. It is certain that he did so on his second expedition a year later and the prolific grain soon spread through much of Europe, Asia, and Africa. By contrast, the Vikings' contact with "self-sown wheat fields," apparently maize, in northeastern North America (about 1000 A.D.) was not followed up.

Other sailors and soldiers under the Spanish flag found Indians and their corn patches throughout much of the western hemisphere. On one trip, Diego Columbus, brother of the great discoverer, traveled eighteen miles through fields of maize in Central America. Conqueror Hernán Cortés came upon corn practically everywhere in Mexico, and the far-ranging Cabeza de Vaca found it under cultivation in Texas.

Terraced grain plots sculptured by Inca tillers high in the Peruvian Andes were observed by Francisco Pizarro and his invaders. Explorers of the tropical Amazon and Orinoco basins and of Florida's subtropical expanses also found the bountiful crop. Gonzalo de Oviedo, a historian who traveled widely in the new domain of the Spanish crown, devoted a chapter to Indian corn in his classic history of the Indies in 1535. And Pedro de Castañeda, the recorder of Francisco de Coronado's expedition, told of finding corn in 1542 as far north as present-day Kansas. These and other Spanish explorers made countless references to maize. Travelers from several European monarchies chanced upon the great Indian grain—the Portuguese in Brazil; Jacques Cartier in the Canada of 1554; Samuel de Champlain seventy-three years later on both sides of the Saint Lawrence River; and French Jesuits and *voyageurs* throughout the Mississippi Valley and up the Missouri into Dakota country.

The greatest Indian civilizations—Maya, Aztec, Nahua, Inca, Pueblo, and Iroquois—were based on an undergirding of corn culture. The Indians did more than domesticate corn. They also developed a number of ingenious tools and techniques for handling and processing it, to which they clung, despite adopting many ways of the European conquerors. In fact, maize culture today among a number of the tribal peoples of Latin America is remarkably similar to the way it was when Columbus and the *conquistadores* arrived. Anglo-America has made the greatest changes in the ways of producing and using corn.

The origins of English America were bound up with Indian corn in all thirteen Atlantic seaboard colonies. But this meeting of East and West actually bore little resemblance to the myths created about it. As a result of Spanish discoveries, the English had known of corn for a century before landing at Jamestown and Plymouth, and their own voyagers had written detailed descriptions of Carolina and Virginia *pegatowr*, or maize. Near Jamestown, the Indians lived in villages and their white neighbors on farms. Throughout wide areas of the Americas, tribes had changed from nomadic to settled or semisettled lifestyles because of corn, bean, and squash growing. The legendary "woods against a stormy sky" along New England's "stern and rockbound" coast had been removed for as much as ten miles inland in some places by Indian agricultural clearing. Primitive, wildernesswise Squanto, the Pautuxet Indian who made corn planters of Pilgrims, had spent several years in England and spoke the language of the whites. And there had been many American thanksgivings by Indians and immigrants long before the little band of Plymouth survivors offered up prayers of thanks in that bleak November of 1621 not for a turkey feast and abundant times, but for a twenty-acre crop of corn and the bare hold on life which it gave.

Sir Walter Raleigh's ill-fated colonizing scheme on the Carolina coast turned up the first on-the-spot English descriptions of Indian corn. In 1584 Arthur Barlow told of "corne which is very white, faire, and well tasted." The Indians, he said, planted their crops in May, June, and July, reaping a stretched-out harvest in July, August, and September. He and Thomas Harriot told of wooden mattocks used for digging and hoeing the soil. The latter explorer likened Indian corn to grains from Turkey and from Africa's Guinea Coast, a tribute to the worldwide seed scattering that followed Columbus' voyages. Harriot noted many colors of maize, white, yellow, red, and blue, and told of making sweet flour and good beer from this grain. The stalks were ten feet tall and each bore up to four "heads," or ears, with five to seven hundred grains to the ear. Another sojourner of Raleigh's company, John White, left a classic sketch of the Indian village Secotan showing three sequential stages of corn growth.

Then came Captain John Smith in 1607 to Jamestown, Virginia, tidewater country and corn country. Smith was a man out of favor, sentenced to die for the loss of two companions, but he put his reputation and the colony back in business, and he did it with Indian corn and Indian aid, particularly the instructions of Tassore and Kemps and the friendship of Pocahontas. Smith's "clever dealings" in securing the precious grain from his dark-skinned hosts of the Powhatan tribe probably saved the colony from quick starvation. The captain's strict work ethic, "he who works not, eats not," forced colonists to plant corn rather than hunt for gold and, despite heavy losses of life, revived their settlement. Using the Indians' seed and methods, these newcomers raised forty acres of maize in their second year. After 1612, the tobacco planting program of John Rolfe provided a cash crop, and from 1616 to 1619, private ownership of land through hundred-acre "headrights" was instituted by Deputy Governor Thomas Dale and Governor George Yeardley at the direction of the parent Virginia Company. In 1619 the first representative assembly in English America, the House of Burgesses, passed laws to insure enough corn plantings for the continued survival of Virginia Colony.

They had come for gold, those haughty, smug adventurers of 1607, and an appalling percentage of them had died. Indian corn provided both food for the fortunate ones and a key to the greatest treasure store of America's history, a harvest mined from virgin soils. Virginia has never had a famine since, and agriculture is still America's greatest industry. So far removed were the perilous first years at Jamestown that by the early 1630s the colony was exporting corn to Caribbean islands and to a place called Massachusetts.

The Pilgrims and other English immigrants of Plymouth, New England, experienced a replay of Jamestown's ugly times—hunger, death of half the original settlers, but eventual salvation for the colony, thanks to Indian help and corn. That first grim winter of 1620–1621 would have spelled doom for the settlement of huddled migrants had it not been for thefts and gifts from native inhabitants. Myles Standish told how the English found "goodly ears" of corn buried in small mounds. These devout pioneers called them "God's good providence" and helped themselves to the fruits of Indian toil and storage. (The local Indians had been all but wiped out by a smallpox epidemic.) They bought, in addition, eight hogsheads (more than fifty bushels) of grain and beans from nearby tribespeople, who taught them how to grind and cook corn. And the hard winter passed.

Stored corn was a short-run means of life support. The storehouse of Indian knowledge would prove more helpful for the long pull. In April of 1621, with lean months and the slow deaths of comrades from hunger and disease fresh in their memories, these Plymouth colonists turned to growing their own food. William Bradford told the story of how they began "to plant ther corne, in which servise Squanto stood them in good stead, showing them both ye manner how to set it, and after how to dress and tend it." Squanto had been captured by Thomas Hunt of Captain John Smith's 1614 expedition. He was sold into slavery in Spain, but reached England, where he learned to speak English. Returned to his native America, he proved indispensable to the survival of Plymouth colony by securing for it the friendship of Pautuxet Chief Massassoit and by serving as an agricultural advisor. A fertilizing technique was part of the lore which Squanto passed on to the Pilgrims. Small, shadlike fish with the unlikely and unaccountable name of *alewives* were netted as they swam upstream in countless millions to spawn. These were planted by the Indians, one or two to a hole, with grains of seed corn. This was not the only area where the first American inhabitants had learned the value of fertilizing. Bird guano had long been used by Indians of Peru to boost their grain yields, and fertilizer from bat caves had been utilized by Hopi and Zuni farmers in New Mexico. Both Indians and Pilgrims added their faith to the fish as they offered up prayers to their respective gods for rich harvests. Whatever the reasons, a crop of twenty acres and a thanksgiving feast followed the labors of the New Englanders, and it seemed that the crisis was past.

But more settlers arrived before another growing season, and the toil for bare survival was repeated through two more summers. A severe dry spell during the green-corn stage of 1623 put these God-fearing souls on their knees for a nine-hour session of prayer. Rains came the next day, and the crop was spared. In that year the colony switched from communal to private land holdings of one acre per family, with the requirement that each family must feed itself. The reported result was a great increase in corn production over the Plymouth Company's former system of common land ownership. Men, women, and children were said to have trudged off to their plots more willingly to reap the harvest of their own labors. As in Virginia, so in New England, the cherished American institution of private land ownership began as an incentive to raise corn.

Soon the deadly hand-to-mouth cycle was past, and Plymouth settlers began "to praise corne as more pretious than silver, and those that had some to spare begane to trade one with another for small things by the quarts, potle, peck, etc., for money they had none." By 1624 they had enough surplus to sail a shalop loaded with corn north to the Kennebec River settlement of Maine, where they traded the cargo for seven hundred pounds of beaver furs. A quarter century later, John Winthrop would complain of corn shortages in Massachusetts because so much had been exported to the Azores and West Indies.

In Maryland, Catholic colony of the Calverts settled in 1634, the natives also made helpful contributions. Indians who were abandoning their village sold ready-made dwellings and cleared maize fields to the settlers. Experiences here and in the later Atlantic Seaboard colonies were less harrowing than in Virginia and Massachusetts.

The English firmly established their beachhead in America thanks to a native grass and friendly guidance by Indian inhabitants. From these coastal points and this perilous time, the account of American expansion to the Great Plains of the trans-Mississippi west was a story of many things. But none was more basic in its shaping influence than Indian corn and the sustenance, the culture, the toil, and the pleasure that it brought to millions of farmers. After Jamestown and Plymouth, corn culture evolved from a bare survival stage into the broader era of pioneer development across the virgin lands to the west.

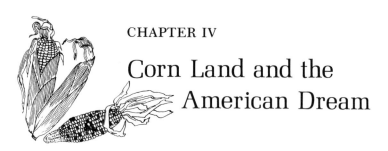

CHAPTER IV

Corn Land and the American Dream

Those who labor in the earth are the chosen people of God, if ever He had a chosen people.

THOMAS JEFFERSON

One can perhaps acquire a feel for the importance of the leading crop in the nation's corn lands by listening to some of the ways the states have characterized themselves. A collage of mottoes, expressions of pride, and lines of music has surfaced from the past and has been sustained in the consciousness of the modern corn heartland. Illinois points to its first ranking among the states and its position ahead of most of the world's nations in corn production. Claiming a close second is neighboring "Ioway," "where the tall corn grows." Nebraska fields its athletic teams of "Cornhuskers" and long staged annual shucking contests, while in Oklahoma, "the corn is as high as an elephant's eye." "Corny as Kansas in August" has been immortalized in song, and "Old Virginny" is "where the cotton and the corn and 'taters grow." Kentucky, with its reputation for blue grass and tobacco, is also a famous corn-whiskey state whose citizens are still referred to as "corncrackers." The "Cracker State," Georgia, is so named for its one-time class of poor white corncrackers, users of hominy blocks to crush the grain. The "Volunteer State," Tennessee, is also the "Hog and Hominy State." The corn palace of South Dakota is known nationwide, and there is no keener competition in any section of the Corn Belt state fairs than in the booths of golden grain. These are but a sampling of the legendary, the lyrical, the folklorish, and the fabled dimensions of a commodity. More than shallow, catchy phrases, the expressions are rooted deep in the lands of America's most important crop.

Across the north-central states from Ohio into Nebraska and South Dakota and from the Ohio River and Ozark Highland north to the

Great Lakes and southern Minnesota lies a flat region of great fertility. This is the American corn belt. It covers an area of over fifty million acres and produces most of the nation's feed and seed corn. The dominant corn-producing area for more than a century, its importance is widely known to most Americans.

It is far less widely known that there have been several belts of American corn lands, each of great importance in its own time. Because of their sparse population, the Indians north of Mexico had no maize belt comparable to modern areas of corn cultivation. Rather their production can best be represented as a number of spatters on the map. The first true corn belt was the Atlantic Coastal Plain from New England to Georgia. Initially contained near the tidewater strip by forests and occasionally hostile Indians, it slowly broadened into the Piedmont, which was the eastern foot of the Appalachian rampart. In time span this coastal corn belt existed throughout the colonial period and well beyond.

Land control and ownership having been the biggest issue in the French and Indian War (1756–1763) and corn having been the use most land was put to, it follows that the belligerents fought in no small part over control of corn lands. Sometimes known as the Seven Years War, this worldwide struggle started in the North American colonies over issues of colonial origin. In a real sense the colonial objectives were both won and temporarily lost in one year (1763). Although France and her Indian allies were defeated in that year, England tried to prevent her land-hungry colonists from crossing into the trans-Appalachian corn and fur country by "drawing" a restraining Proclamation Line down the crest of the mountain chain and forbidding settlers to cross it. Designed to prevent costly disputes with Indians, the Proclamation Line threatened to cut pioneer families and speculators out of the very corn lands they had fought for, but they violated the restriction in large numbers. The Decade of Controversy, prior to the Revolution, began and ended largely on questions of western land, from the coercive Proclamation Line of 1763 to the restrictive Quebec Act of 1774.

Every American war from the French and Indian through the Civil War opened new lands for farmers to settle, sometimes by treaty, sometimes by pacifying the Indians or by contributing to improvements in transportation or changes in land laws. The Peace of Paris of 1783, which ended the American Revolution, created a new nation,

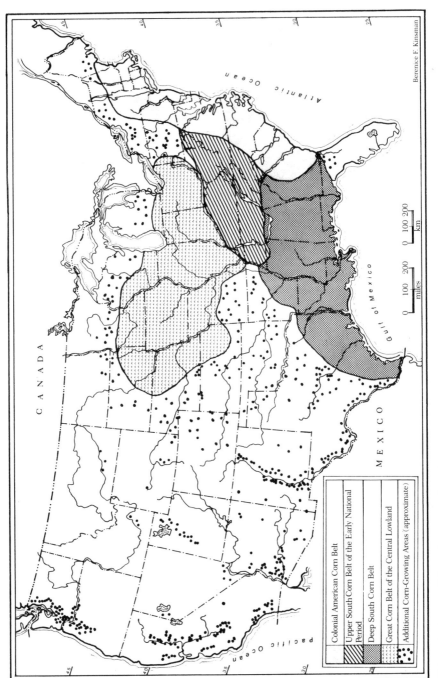

Historical map of American corn farming.

fixed its western boundary at the Mississippi River far beyond the hated and outdated Proclamation Line, and thus gave the United States title to its greatest future "corn patch."

The Articles of Confederation period began in 1781 only after a long delay over cession of western lands to the national government and ended, for practical purposes, with the Northwest Ordinance of 1787, a landmark law in the truest sense of the term. Trans-Appalachian pioneers cleared and planted a big new corn belt westward into Kentucky and Tennessee beginning in the 1770s and 1780s. Encompassing neighboring Virginia, this belt held its first-rank position almost to the middle of the nineteenth century. As late as 1840, Tennessee was the leading corn producing state in the nation, with Kentucky ranking second and Virginia third. (Virginia then included West Virginia.)

During the early 1800s two more corn belts began to stretch steadily westward, although they were resisted by Spanish, Indians, and English from the Gulf of Mexico to the Great Lakes. Thomas Jefferson, always eager to support small farmers, greatly increased the corn lands and doubled the nation's area with the Louisiana Purchase of 1803. By a reasonably successful conclusion of the War of 1812, the Americans brushed aside most remaining Spanish, English, and Indian barriers and swung open the western door which had been left ajar in the Treaty of 1783.

These two new corn belts now surged west. The northernmost one extended from Ohio to an area beyond the Mississippi and became the great American Corn Belt of modern times. From Georgia the southern belt stretched along the Gulf Plains into Louisiana. This latter corn corridor overlapped the cotton belt. Although corn has not been given the diplomatic and literary recognition of "King Cotton," its acreage in the combined southern states surpassed the fiber crop by about three to one throughout the pioneer period. In a typical year the value of the American corn crop has been more than twice that of the cotton crop, and today corn acreage is more than five times that of cotton. A gang of slaves could grow more cotton than they could possibly pick, and it was an economy of time to direct part of their energies to raising the principal human and animal food—corn. In Georgia, for example, it became the practice to plant ten acres each of corn and cotton per slave.

The older eastern states watched westward expansion with mixed feelings. It meant new raw materials and markets, but it also shifted the nation's political power and center of gravity toward the west as new states entered the Union and new representatives took seats in the nation's capitol.

There are, and long have been, millions of acres of corn outside these great belts in patches from coast to coast and border to border. Indians grew the crop successfully in many other areas, including the desert southwest, the northern Great Plains, and in the last corn frontier of the white migration, Oklahoma, which was occupied by "Sooners" in 1889. Mormon settlers tended prosperous irrigated corn patches in Utah. The modern Far West grows significant amounts of corn. And countless home gardeners all over the nation raise sweet corn, which, if it could be figured into the bushel and ton totals of production, would push corn's statistics well ahead of the modern-day three-to-one ratio over wheat, since the latter grain is not garden-grown at all. Sweet corn occupies more acreage than any other vegetable grown for human consumption in America.

A long list of treaties, most of them with Indian tribes, opened vast acreages of potential western corn lands to settlement: Fort Stanwix, Hard Labor, Lochaber, Sycamore Shoals, and Greenville, to cite but a few. In the northern march and Fallen Timbers battle leading to the Greenville Treaty (1795), U.S. General "Mad Anthony" Wayne's scorched earth policy destroyed Indian maize fields for fifty miles along the Maumee River in Ohio. Then there were agreements such as the Treaty of Saint Louis (1825), removing northern Indians westward, and a series of "ultimatum" treaties that uprooted the Civilized Tribes of the South and sent them along the tragic "Trail of Tears" across the Mississippi. The "Corn Treaty" was signed June 30, 1831. By its terms, Chief Black Hawk gave up dealings with the British, consented to the building of government posts and roads in the Illinois area held by his people, the Sauk and Fox, and agreed to accept payment for the corn which his tribe was obliged to abandon. This treaty was broken by whites and the Black Hawk War resulted when Black Hawk's Indians attempted to procure supplies of corn.

Such was the fate of most Indian treaties. "Forever" again and again turned out to be but a short time; the democratic process itself put pressure on the government, through whites who could vote and at the expense of Indians who could not, to renegotiate agreements and nudge the outnumbered first inhabitants ever westward so that settlers could claim their lands. The Manifest Destiny outbreak in the 1840s added Texas, Oregon, California, and the remainder of the land west of Louisiana Territory to the nation. There was much future corn acreage in these acquisitions.

Wars and treaties affected corn lands in more ways than the enlargement of the nation. Land-rich and tax-poor states, such as Georgia, North Carolina, and Virginia, often paid their military veterans in land. Grants varied by rank of the recipients. A private in the Revolutionary War from North Carolina could receive 640 acres while a general was eligible for over 12,000 acres.

Any crop that can be grown throughout such a large land as the United States has a wide range of climatic and soil tolerances, and corn is such a crop. Since it can grow at high elevations and high latitudes, there is only a small area of the nation that is out of reach of this adaptive grain. Tolerant of wide variations in temperature and growing season, it thrives from the tropics to points north of the United States–Canadian border. Rainfall has posed limitations only in the very arid regions of the country, and in many of these, Indians, Spanish padres, Mormons, and modern farmers have brought irrigation waters to the rescue. Josiah Gregg observed in the 1840s that such agriculture in New Mexico was confined to valleys of "constant flowing streams," where irrigation was much more dependable than rain. Hernando de Soto's expedition of the early 1540s reported that in parts of the Great Plains the Indians did not grow corn because there were too many buffalo to permit protection of the crops. With the exceptions of this area, the Pacific Coast, the Canadian border region, and some of the seemingly unbreakable sods of the interior, the Indians had introduced corn to practically all of the areas of the country where it is grown today. Recognizing that they had also developed all the major subspecies and colors of the grain, one can only marvel at their contribution to corn culture.

Soils are scarcely more limiting to maize than weather and climate. Scandinavian visitor Pehr Kalm's report in 1750 noted that sandy loams were best. Too much clay caused poor development of kernels and late ripening. New Jersey, he said, was so sandy that it would have been hard to live there without corn. The plant's deep roots reached far into porous soils for the scarce nutrients. Kalm went on to observe that corn, of all the crops, provided best usage of the driest and poorest soils. Indian experiences in the sandy southwest seem to have confirmed his observations.

The Swedish scientist was probably on less firm ground in his statement that very rich soils caused a growth of too much stalk and leaf and not enough grain. This was more characteristic of wheat than corn. Pioneer farmers sought out the river valley soils because of their great fertility and the accessibility of cheap transportation via river-boat or barge. Bottomlands, the floodplains or first level of soils above the riverbanks, were very rich and much better adapted to corn than cotton, for example. Settlers along such streams as the Ohio, Mississippi, Missouri, Tennessee, Cumberland, and smaller streams of the Gulf Plains could lose one corn crop of every three from flood damage and still continue farming the land successfully owing to renewed richness from the alluvium. (This, of course, was before the modern era of scientific fertilization.) Records of bottomland farmers show that they paid a higher price in corn lost to flood waters than in any other crop.

The so-called second bottoms, leveled by meandering streams in the geologic past but high enough to be free of most floods, were also made up of fertile alluvial soils combined with decomposed organic matter. However, these soils near streams had a serious drawback in addition to flooding. It was widely believed that the lowlands generated bad air, particularly at night, and that the bad air caused disease. (*Malaria* is derived from the Italian words for bad air.) Therefore, these lands were often avoided even though they were quite fertile. Later generations would discover that the real culprits were germ-carrying anopheles mosquitoes, which plagued most lowlands.

Other instincts of the strongly tradition-conscious settlers about the soil were also to prove wrong. From two centuries of experience in the eastern woodlands, farmers had come to believe that land that would not grow trees would not produce corn or other crops well either. Thus some of the treeless prairies of northern Indiana and Illinois and southern Michigan and Wisconsin went untilled for years, as did wide areas of the trans-Mississippi West. The "Great American Desert" myth as it applied to the Great Plains grew out of this false notion. Migrants to the Pacific Coast in the 1840s and 1850s preferred to clear and farm the timbered lands first, temporarily passing over such fertile soil as Oregon's Tualatin Plain. Time would prove them wrong. Not poor soils, but low rainfall, high winds, fires, and ancient clearing by Indians were the reasons for the lack of timber in many areas. When their superstitions were at last overcome, the settlers literally stormed onto the prairie lands.

Corn was a mover in early America. Not only did it advance the economy greatly, but it also lured people—millions of them—westward. Manifest Destiny, the urge to the west in the 1840s, was sparked partly by land hunger, which was usually an appetite for corn land. That creator of frontier ideas or hypotheses, Frederick Jackson Turner, probably overemphasized *free* land as a westering motive, for land was by no means always free, but he did not greatly overstress land itself.

If not always free before the Homestead Law of 1862, land could certainly be bought at much lower cost in America than in contemporary Europe or in the United States of modern times. European traveler William Faux wrote in 1819, "Nothing is reckoned for land [in America]; land is nothing; labor is everything."

The close ties between land titles and corn crops is shown by "corn patch and cabin rights" in Virginia in revolutionary times. Novelist Mary Johnston was right. In the Old Dominion, the planting of Indian corn was "prescribed by law" as a matter of survival. The Virginia State Assembly's first land law (1776), which governed free public land in the back country, provided that "no family shall be entitled to the allowance granted by this act unless they have made a crop of corn in that country, or resided there at least one year since the time of their settlement." Three years later Virginia passed a land law offering four hundred acres of land in Kentucky Territory for a total of ten dollars (two and one-half cents per acre) to any householder who would settle the tract, build a cabin, and plant a corn crop within one year. A number of early laws restricted exports. These and other such measures were on the law books of Virginia, Maryland, and other colonies and states for many years. Often agreements were very informal. Lands were bought and sold on the basis of no firmer claim than the fact that a settler had raised a crop of corn, for it was universally understood that, except for garden vegetables such as the very important turnip, corn would be the first crop.

One corn and hog farmer in 1784 acquired an entire section of prime land—640 acres, or one square mile—in the Cumberland Valley of western North Carolina, which would become the State of Tennessee twelve years later. The price he paid was "the difference in value between two saddle horses." (Today farm land of that quality might sell for $2500 to $3000 per acre!) A succession of national land laws from 1785 to 1820 provided for the sale of public lands at prices from one to two dollars per acre. The minimum size of plot varied from two square miles (1796) to eighty acres (for a total of $100 cash) under the Land Law of 1820. Preemption legislation in 1841 enabled farmers to become legal squatters by settling on government-owned land and later paying the going price for it or packing up and moving when the government surveyors arrived. This procedure gave them a chance to raise enough from the plot to pay for it before it was surveyed. Here and there a settler might be able to afford the price for the place of a farmer who was selling out or might buy a cabin built by indentured servants of a western land speculator, such as George Washington or Christopher Gist. But at the risk of a tomahawking, pioneers could continue to farm free land if they were willing to move often.

And indeed, they moved frequently, although for more reasons than the appeal of low-priced land. Except for the Pennsylvania Dutch (Germans who began coming to the Quaker Colony about 1700) and their descendants, Americans were generally heedless of the need to take care of their soils. Corn was hard on the land, especially when planted year after year on the same patches with no rotation and little or no fertilization. Pioneers could only afford this wasteful practice because land was plentiful and cheap, and it was less expensive to clear than to fertilize. When their corn acres became tired, they would clear adjoining timber or perhaps move on. Tired soils doubtless served as many eviction notices as mortgage holders in pioneer America. Squatter's rights before and after the preemption law beckoned settlers to put out the fire, call the dogs, count the youngsters, and point their wagons or flatboats toward new homes in the West. (Not all corn farmers were movers, however. Some Indians of the Southwest still grow corn on the same lands and with about the same methods as did their ancestors more than 150 years ago.)

The urge to move was often accompanied by a desire for larger holdings. Close neighbors were a social blessing but an economic curse for the many small farmers who had dreams of becoming country squires. By moving they might widen the horizons of their estates. A nineteenth-century observer of McLean County, Illinois, captured the picture well in sharp and critical poetic lines.

> Early on hand in the dewey morn,
> They turned the soil and they raised the corn;
> They gathered the corn and they raised the pork,
> They steadily prospered by constant work,
> They sold the pork; with cash in hand,
> Their instant thought was "get more land."
> They got more land, and they raised more hogs;

> They got more cash, and the greedy dogs,
> In ceaseless grasp and thankless work,
> Purchased the chase—more corn, more pork,
> More cash, more land, more corn, and then
> More land, more cash for greedy men.

What irony that these lines still seem entirely applicable to the mechanized farmer's present-day chase for more land, more cash, more expensive machinery.

Unhappily, these tillers of the soil were also killers of the soil, unaware of the necessity to conserve that great natural resource. Even as wise a planter as Thomas Jefferson thought that the supply of American soils was inexhaustible. Encouraged by the habit of becoming a free squatter, then moving on if necessary, the average settler treated the soil like a mine from which he would quickly extract the mineral wealth by overcropping. The fire-clearing of timber in the first year or two and of stubble and stalk thereafter was damaging to the soil. Thomas Harriot reported in the 1580s that the Indians did not scatter the ashes evenly over the soil which they had cleared by burning. The limited value derived from ashes was therefore spotty.

Another problem was the pioneers' habit of plowing and cultivating in whatever direction it was convenient to turn a furrow or turn a team around. Uphill-and-downhill tilling made much land useless for it encouraged the erosion of great gullies. English agriculturalist Hugh Willoughby, after viewing these practices in the hills of eastern Kentucky in the twentieth century wrote: "Never have I seen such astonishing examples of 'vertical agriculture' as in this area. I have seen corn grown on a slope I could not climb up without using one hand at times." Washes from tilling up and downhill were especially bad in the Piedmont belt on the eastern edge of the Appalachian cordillera. Farmers dumped corn fodder crossways in these "runs" as they were called in some areas, then shoveled dirt on top to hold it down in an attempt to stop the washing away of the soil.

But more was needed than stopgap measures. Contour plowing and planting had long been known and were practiced by the Pennsylvania Dutch, whose soils today are nearly as well preserved as when first cleared and plowed 275 years ago. The principle was simple enough: plow and cultivate around instead of up and down the slopes so that runoff moisture would be forced to seep across furrows rather than gush down them. In this way the runoff would not gain the momentum to wash out gullies. Some Indians had learned to plant on the contour, and in a sense, the terraced strip fields on Andean hillsides in South America were designed on this principle. Many American farmers thought the curved furrows and corn rows of the scientific contour tillers were plowed and planted by drunks who could not drive teams in a straight line. Those same guffawing settlers heedlessly shipped tons of precious topsoil seaward with every hill tilling and every freshet and flood.

Numerous crops were involved in rotation cycles, but corn, as the leading product, was the common denominator. Crop rotation was an old practice but was not used widely before the mid–nineteenth century. As planter Thomas Jefferson was aware, rotation of corn, wheat, clover, timothy or other hay, and beans greatly improved yields. Cowpeas, after their introduction from Africa in the early eighteenth century, became very valuable in crop rotation. These and other legumes added nitrogen to the soil. With a number of fields, a settler could grow all these crops at the same time, changing from year to year on each plot. Crop rotation was also helpful in disease control. For centuries the Indians had practiced what amounted to "simultaneous rotation," planting beans among their corn.

Fertilization before the late nineteenth century was, like contour farming and crop rotation, only vaguely understood and spottily practiced. Turning—that is, turning weeds, grass, leaves, and cornstalks under by careful plowing, thereby returning humus to the soil—was superior to burning. The Indian and pioneer practice of planting fish tripled corn yields in some instances. Some pioneers put handfuls of manure in holes with the seeds. This was helpful but not as effective as it might seem because the corn roots grew so far that they overshot the concentrated blobs of fertilizer.

There were many kinds of fertilizers and numerous ways to fertilize. "Hogging down" corn in the fields served the dual purposes of making the harvest easy and manuring the soil. Some Peruvian Indians had used both human and animal excrement for fertilizing corn in pre-Columbian times, and the Zuni and Hopi people of the New Mexico–Arizona area enriched their maize lands with bat guano from caves. Well-known and well-to-do American experimental farmers of the early nineteenth century, such as Henry Clay, Elkanah Watson, Edmund Ruffin, John Hardeman, and Thomas A. Smith, knew and practiced manure fertilization and crop rotation. By mid-century there were some chemical fertilizers on the market, and farmers were slowly becoming acquainted with such terms as *gypsum* (sulphate of lime), *bone dust, nitrate of potash, sulphate of ammonia,* and *phosphate of ammonia.* Speedy clipper ships, which had been built in such great numbers that they could not find enough business even in the booming eastern trade with gold rush California, hauled bird guano from the west coast of South America to the United States to be scattered on impoverished soils.

Most agrarians lacked either the knowledge or the desire to haul their rich and abundant livestock and poultry manure to the fields. This has been called the "greatest negligence" in their entire range of agricultural practices. Farmers simply moved to new lands when old soils failed them, and later abandoned farms for the growing cities when good soils became scarce and mechanization created farm unemployment in the latter half of the nineteenth century.

Few were the farmers who heeded the advice from almanac writers of the early 1800s: take mud from your bogs and ponds, and scatter it in the barnyard, dry and pulverize the mix of muck and manure and spread it evenly over the fields. Mix the dung of cattle, hogs, and horses before scattering. The *Old Farmer's Almanack* of 1802 admonished tillers that manure was like money; it was no good until spread. Barnyard manure was "one of the most effective fertilizers for corn," farmers were frequently told. Yet at the Gardner Randolph farm, Randolph's Grove, Illinois, in 1825 manure became so deeply piled around the stock barn that, instead of hauling it off to fertilize their fields, the family dismantled the barn and rebuilt it away from the dung heap. Perhaps they shared the feeling of the many settlers who shunned food grown on manured land.

In time the picture changed, elements of scientific agriculture began to break through and tillers of tiring soils began to understand that barnyard waste could come back to them as bountiful harvests. Governments and colleges and a few inquiring experimental growers were helpful in slowly turning the tide. Newspapers, such as the *National Intelligencer,* Niles *Weekly Register,* and the *Missouri Intelligencer,* and journals, following the lead of the *American Farmer* (published at Baltimore beginning in 1819), supplemented the numerous farm almanacs, providing many tips on farming improvements for those who could read. These new organs also published each other's letters from gentlemen farmers. Thus farm information, and much misinformation, circulated rapidly. In 1862 the National Bureau of Agriculture was founded. Its offspring became the United States Department of Agriculture, which put out great quantities of literature on how to increase farm production. The Federal Patent Office, as early as the 1850s, published detailed annual reports and distributed improved varieties of seed, including large quantities of corn. Vermont Representative Justin H. Morrill sponsored the Morrill Act of 1862, which gave much federal aid through land grants to state agricultural colleges. Their educational impact was vital in modernizing agriculture. The Hatch Act of 1887 established agricultural experiment stations at these land-grant colleges, furthering knowledge of soil and animal science. Still, conservation of land was a strange concept to many settlers. One Ozark foothill farmer thought the word was "corn-servation!"

No statement on improved use of corn lands would be complete without crediting the great impact of Seaman Knapp. Beginning late in the nineteenth century, Knapp launched a program to encourage southern farmers—boys and girls as well as adults—to diversify their farming, conserve and fertilize soils, rotate crops, select improved seed, and organize corn clubs. The results were little short of miraculous. Abandoned corn lands and exhausted soils producing as little as ten bushels to the acre came to life and began turning out from four to ten times the old yield. Organizers formed state and national corn clubs and gave prizes to the best producers. Jerry Moore, a South Carolina youth, raised 228 bushels per acre. Shortly before this the state average had been only ten bushels per acre! The South, which produced one-fifth of the nation's corn in 1890, had raised that figure to one-third by 1912. At the time of Knapp's death in 1911, his corn clubs were in every southern state and many outside the South.

Other corn missionaries, such as Professor Holden and his corn train and W. H. Smith, a Knapp protegé, also took up the cause. Four-H clubs and Future Farmers of America groups spread widely through the country. Such organized programs to train a new generation of scientific farmers became part of the curricula of thousands of high schools and many colleges. The corn lands produced as never before and met the food needs of a fast-growing population.

In spite of all their abuse of soils, Americans have developed and clung to a respect, almost a reverence for land and land ownership. The history of American worship of property, love of the land, the work ethic, and upward social and economic mobility could be written largely around corn. Many immigrants to the new country became land owners for the first time in their lives. They and their descendants took great pride in what they acquired and built with their hands and the sweat of their brows. Migration to America was based in no small part on the life of relative abundance that corn made possible.

As will be seen, migrants to new lands in the West usually tried to arrive early enough in the spring so that they would not miss a corn crop. Often this meant traveling in the cold of winter and early spring. They also had to build shelters for protection against Indians and weather. But there were times when the need for putting in crops was too immediate for delay. Olof Olsson, Swedish immigrant to Kansas in 1869, wrote, "The life of the pioneer is truly difficult. . . . His work wagon is for a long time during the first year his hotel, kitchen, salon, bedroom, and church."

With time and toil, the corn-farm family would change its new estate from camp to comfortable home: first a small cabin, a barn, and a corn crib, then the cabin sprouted wings or additions. "Add-a-room," a thriving modern business, is nothing new in America. Eventually the newcomers might build fences, cellar, pantry, springhouse, icehouse, well sweep and wellhouse, buttery, and smokehouse onto the cabin or close by. In time, too, the sacred feelings toward the land would take on new and saddened dimensions as family grave plots were established. Small sections were fenced off, and stillborns, patriarchs, matriarchs, and children (among whom death rates were very high), were buried in the very lands that had been wrested from primitive forest to productive farm and had grown corn or wheat or cotton. For some pioneer types, family graveyards were added reason to stay on the same farm, while others moved farther west to leave the sad reminders behind.

Land and the accumulation of landed property were at the heart of the great American dream. English philosopher John Locke's ideal of the enjoyment of property was much more fully realized in America than in Europe. It was so not only for farmers but for most town and city dwellers who longed for a place in the country, and for the successful bankers, traders, and industrialists, who almost routinely spent some of their profits on verdant rural estates.

Strong sentiments about the virtue of agrarian life still exist. There are, for example, large government subsidies to farmers (mostly to holders of big farms). Even the school calendar that allows for summer recesses is a reminder of the days when children needed to be free to help their parents on the farm. There was a formal philosophy in addition to Locke's to support the superiority of this agrarian model of life. Physiocrats, as its believers were called, such as François Quesnay of France and Thomas Jefferson of Virginia, believed that land was the source of all true wealth. This concept was related to Jefferson's image of the small farmer as the ideal; that there was something about working the land and bringing forth its increase that breathed the spirit of true democracy and virtue into the very souls of the tillers of the soil. Jefferson's Louisiana Purchase and his Virginia statutes abolishing the last remnants of primogeniture and entail gave tangible support to his philosophical views.

A New England almanac of 1849 had some words "In Praise of Agriculture" which reflected popularly held notions of the day. "There is no class of men, if times are but tolerably good, that enjoy themselves so highly as farmers. They are little kings. Their concerns are not huddled into a corner as those of the tradesman." Another typical sentiment of the same time and area comes from a poem, "The Tillers of the Soil."

> His stacks are seen on every side,
> His barns are filled with grain;
> Though others hail not fortune's tide,
> He labors not in vain.
> The land gives up its rich increase,
> The sweet reward of toil;
> And blest with happiness and peace
> Is the tiller of the soil.

Idyllic as these expressions sound, the work was hard, the hours were long, and for those many farmers who pioneered in virgin soils, there was much to be done before the land gave up its "rich increase."

CHAPTER V

Corn Was Work —and Fun

It is the fashion with AMERICAN farmers to call the husking a "frolic." The cunning fellows know, that if they were to call dancing *work*, it would be a pretty hard matter to get a party together.

WILLIAM COBBETT, 1828

"A lazy man's crop," said a Swedish visitor in the 1740s about American corn because of its high yields per acre. Scotchman John Muir, a century later, disagreed. Before he turned mountaineer, this Wisconsin corn farmer had words with feeling about a typical summer work day. "All together a hard, sweaty day of about sixteen or seventeen hours. Think of that, ye blessed eight-hour day laborers!" Farm work hours varied little from tidewater to frontier.

Rare was the farm task that was done without hard work, and corn proved the rule rather than the exception. Back-breaking, sweat-soaking, hand-callousing jobs characterized every phase of the plant's culture from clearing, plowing, planting, and cultivating to cutting, shocking, shucking, hauling, shelling, grinding, and worst of all, tramping silage. Yet for all the drudgery, few aspects of rural labor were more mixed with a hint or a pile of pleasure than was corn culture. There were many uses of the product purely as games, but often as not, farmers found fun in their work. In the case of the husking bee, they went so far as to unconsciously institutionalize it.

The literature of pioneer America is rife with signs of the work ethic. "Awake, O Ichabod, Awake!" was an oft-repeated call to farm chores from the *Old Farmers' Almanac*. Ben Franklin's exhortations were as frequent as they were pungent, and from Puritan writers to *McGuffey Readers*, the message was the same. "America is the land of work," and all things come to those who labor. A literate farmer in 1800 advised the tiller of the soil to "rise early and to be in the field at least half an hour before the sun visits the eastern hills; by this means he may perform half his daily labor before the heat of the day, and thereby may rest in the heat of it." Many farm workers were on the job while the morning star hung brightly over the eastern horizon.

Whatever the value of these words on pioneer farm labor, the pronoun *he* was at the least misleading, and at most grossly inaccurate. Historical accounts of rural America have scarcely begun to credit women for the large share of outdoor farm labor which they did. Men were usually absent from the fields for many days each year, taking part in military or Indian campaigns or political assemblies, trading, salt boiling, hunting, trapping, scouting for new home sites, or else taken off the work force by illness, injury, or death. At such times women, with the help of children, often bore the entire load of farm drudgery, and they carried a sizable share of the responsibilities when the men were on hand. While women in all regions of the country did prodigious amounts of work, those in the West and those who were first-generation immigrants were more likely to be found laboring in the fields and forests.

Journalist-essayist Charles Dudley Warner had words of appreciation for the unheralded heroes of farm work. "After everybody else is through, [the farm boy] has to finish up. His work is like a woman's— perpetually waiting on others." Corn, from the standpoint of labor, was a pioneer women's and children's crop just as it was men's work. A United States Agricultural Patent Office official, writing at the beginning of the Civil War, strongly advised farmers not to invest in the newfangled corn planting machinery, since the job could be handled much cheaper and as efficiently by handwork of the women and children. The fact that corn was the one grain crop that could be successfully produced on a large scale by hand helps to explain why the Indians grew it in such quantities without the aid of draft animals and why women and children figured so largely in the work force.

From the New England of 1809 came a strong dissent against sending women to the fields to work "like Amazons, with pitchfork and rake. . . . T'is abominable!" The very term *sending*, however, implied a subordinate role for women—a role of obedience rather than equality. And on some farms, notably in the back country, wives and daughters were forced into roles little different from slavery. William Byrd wrote pointedly about the laziness of men and abuse of women in the North Carolina–Virginia corn frontier of 1728.

Indian corn is of so great increase, that a little Pains will subsist a very large Family with Bread. . . . The Men, for their Parts, just like the Indians, impose all the Work upon the poor Women. They make their Wives rise out of their Beds early in the Morning, at the same time that they Lye and Snore, till the Sun has run one third of his course, and disperst all the unwholesome Damps. . . . When the weather is mild, they stand leaning with both their arms upon the corn-field fence and gravely consider whether they had best go and take a Small Heat at the Hough: but generally find reasons to put it off until another time.

Over a span of several thousand years, farm tools and work had changed very little. A basic characteristic of tilling the soil was its simplicity. Except for a few tasks such as heavy plowing, children by the age of ten or twelve years had learned and could reenact the entire cycle of pioneer farming down to the smallest detail, and a ten-year-old could shuck corn "almost as well as a man." Life was so simple, in fact, that each generation lived very much as the previous one, and as a European visitor observed, "*the rich and the poor all fare much alike.*" Rural society was thus stable, and even static, despite much diversity of labor institutions and systems.

Among institutions, corn-related labor systems in America extended over the whole range of those known to the world. This grain's culture occupied a much larger part of the work time of a higher percentage of people in pre–twentieth-century America than any other crop or craft. Not only was it the chief farm product, it also required a large amount of individual labor. Cotton and tobacco were regionally too limited to be in contention. Nor could wheat compare since although it required much effort to harvest, it took fewer hands to grow because, once it was sown or drilled, it was never tilled. And manufacture and trade, overall, involved a much smaller part of the population than did agriculture. As the chief labor commodity, corn had a great impact on patterns of American work.

Communal labor systems probably characterized most corn raising in America before the early 1600s. Indian land use was primarily cooperative within tribal or subtribal units or feudal in some areas of Mexico and South America. These communal conditions remained among the Indians of Anglo-America until they were wiped out by white invasions, and in fact, there are still some communal cornfields among the Navajo of the Southwest. Although Indian land ownership was communal, there was some private working of the soil. John Bartram, traveling near the Little Tennessee River in the mid-1700s, observed the "little plantations" of the Indians, "corn, beans, etc., divided from each other by narrow strips or borders of grass, which marked the bounds of each one's property, their habitation standing in the midst."

It has been noted that at Jamestown before 1616, and in the Plymouth area prior to 1623, the joint stock company farm structure was also communal. Then the company directors provided for highly successful, privately owned, single-family farms that came to dominate the agricultural ownership and labor scene in most of English Colonial America and the United States. True, colonies such as Robert Owen's New Harmony, Indiana, and the Oneida Colony of New York were communal. However, owing probably to the entrenchment of private land holding as an institution and the ready access to an abundance of land, these communes did not prove highly successful. The Amana Colony, an Iowa commune that grew much corn, was an exception. It succeeded both before and after it went capitalist in 1932.

Family farm units shaped American character. Without question they fostered self-reliance as children learned much of the discipline of work from chores related to Indian corn. At the same time they felt a measure of security and personal worth from being wanted and needed to sustain the family. In bad seasons there was all the more need for family teamwork. "Poverty on the farm builds character," said an observer of American life in the 1840s, while "poverty in the city breeds vice." Several decades earlier, an English traveler in America, Frances Trollope, observed what she called the "nearly complete independence" of forest farmers who raised Indian corn.

Yet along with the independence and individualism, young pioneers learned something of the spirit of cooperation—family member to family member, and neighbor to neighbor. Individualism and cooperativeness went hand in hand. When a work animal got sick or died, a neighbor would usually lend a hand or a horse. A spring plot unplowed was a winter season unprepared for. However, this family teamwork and exchange of chores with neighbors, perhaps paradoxically, did not seem to transfer to the larger communal enterprises.

Not all about corn work was free and individualized and voluntary sharing. Bound labor abounded in the New World both before and after Columbus. Feudal serfs had tilled maize fields of Mayan, Aztec, and Inca overlords for hundreds of years. The western Europeans stamped out serfdom on both sides of the Atlantic by abolishing feudalism in Europe and toppling Indian power structures in America. Yet ironically, they soon recreated feudal labor systems of their own in the Western Hemisphere. The Spanish had their *encomiendas*, England its *proprietorships*, the Dutch a *patroon* system, and France the *seigneurial* structure in America. Owing to an abundance of land and inability to hold laborers, most of these systems soon collapsed.

But there were other schemes to replace them. Dating from the early years of English colonization, indentured servants sailed west from Europe to America. There they worked as debt servants on farms or plantations or at crafts for three, five, or seven years, and sometimes longer, in exchange for passage to the "Promised Land." Perhaps as many as half of the working-age white immigrants to England's seaboard colonies came as indentured servants. Debt servitude also prevailed in the maize plots of Spanish and Mexican Texas and New Mexico in the form of peonage, which continued to some degree until 1867.

Black laborers arrived in Virginia Colony by 1619, and although they were probably employed first as indentured servants, slavery evolved within a few decades. Initially it seemed desirable to keep these trained laborers on the job for longer and longer terms rather than to instruct newcomers in often complex work. Then it was considered necessary, for vague and little-understood socioeconomic reasons, to keep them as slaves, perhaps because they had been kept as slaves. Plantation owners and overseers could more easily identify and recover black runaways than whites. Unlike indentured servants, the slaves received no deliberate instruction for freedom. On the contrary, their masters systematically denied them the education, skills, and anything else that might have helped them escape from bondage, even including their African drums.

The freedom training that the slaves got was accidental and, like so much of their toil, involved corn. Beyond the work for their masters, which occupied much of their laboring time, many slaves were given corn plots to work on their own. Planters provided these patches for the subsistence of the slaves and their families in order to cut down the costs of feeding the work force. This scheme worked better with the task system of labor, whereby hard-working slaves could earn free time for themselves, than with the less common gang system. In planning their own schedules of preparing the land, planting, cultivating, and harvesting corn on these "freedom plots," and in feeding their hogs, cattle, and poultry, the slaves acquired valuable training as free farmers, small-scale agricultural entrepreneurs. This experience was of great importance in the transition from slavery to freedom in the 1860s, '70s, and '80s, when, as has been often observed, the first interest of the freedmen was "a place to plant corn."

Tenancy as a means of corn cropping was almost as old as capitalism in America. Particularly in the post–Civil War era, large landowners applied the share-tenant, crop-lien, and sharecrop systems to various crops, notably cotton and corn. Sharecropping is still common in America. Employment of many hired hands in a "factory-in-the-field" arrangement has long been used by holders of large estates. Finally, owners of mechanized contraptions have for many decades engaged in job-contract labor, whereby they move from farm to farm like an army of locusts, spraying, harvesting, or performing whatever tasks they are equipped and paid to do.

While Indian corn has been often related to the bound or semibound labor of men and women, serfs, indentured servants, slaves, impoverished tenants or sharers, even children bound to parents or loaned by the "workhouse for orphans," it has also had an egalitarian effect. Because of its high yield per acre, millions of independent small farmers have been able to survive even though they were too poor to afford more than a few acres of land each. Scarcity and high costs of labor in early America also contributed to a single-family subsistence-farming pattern when bound labor was not available or acceptable.

The drudgery necessary for corn production was fun partly because of the human capacity and need to find a lighter side to toil. The Volga boatmen sing as they heave at their oars; Indian women pound tortillas to the rhythm and tune of their native music; plowhands croon along the toilsome furrow; and farm children race to see which of them can hoe first to the end of the long, hot row and the cool jug with its corncob stopper in the shade of the old white oak!

No more typical example of combining work and pleasure can be drawn from America's past than the husking bee. Perhaps this "joy that welled up in all hearts during the harvest" was a universal characteristic from Greece to Egypt to Rome, from the biblical fields of Boaz to the merriment in old English and New World Indian farms at the reaping time. Pioneer farm folk, by the very nature of their livelihood, dwelled a "fur piece" from their neighbors, and gregarious and lonesome as many of them were, women and men and children alike used every imaginable excuse to get together—weddings, religious camp meetings, logrollings, house-raisings, apple butter and maple sugar boilings, dung frolics (fertilizing of fields), even funerals. For a pleasant way to tackle your farm chores "assemble your army of men, women, girls, boys, and horses or oxen," advised an English farmer in America, and so it was with corn husking or shucking, a task that, as the Indians had long before discovered, was well adapted to concentrated group labor combined with fun. Many a farmer came to a funeral to broadcast news of a corn shucking. Although the terms "husking" and "shucking" were well understood in all regions as being interchangeable, "husking" was the word more likely to be used in New England, while the South and Missouri preferred "shucking." Both terms were used in New York, Pennsylvania, and the Old Northwest.

Husking bees, or frolics, because they happened every year with the fall harvest and on countless farms, were by sheer numbers the most popular social events of the American farm, if not of all America, at least until the mid–nineteenth century. Not that pioneer farmers lacked other amusements. Varying somewhat as to time and place, there were "peanuckle," backgammon, parchesi, euchre, checkers, and occasionally croquet. Grange picnics, chautauquas, fairs, hog callings, shooting contests, horse racing, ice cream festivals, and candy pulls added to the diversions, but corn shuckings outstripped them all. Each family might attend many such work-play activities every autumn. There were variations in the pattern from time to time and from one region to another, but participants generally followed certain basic rules of the game.

Picture a moonlit autumn night, with the chill broken by crackling pine-knot fires kindled with corn cobs atop three or four earth-covered scaffolds. Inside the flickering circle next to the corn crib or granary, or perhaps on the barn floor, were two evenly divided piles of "snapped" corn ears "in the shuck." These were separated by a rail. Two leaders picked teams of huskers who had gathered—perhaps to the number of fifty to seventy-five—from surrounding farms or plantations. The teams usually included both sexes. In the South, slaves and children were often allowed to take part, although in some areas, if slaves were invited, white women were not. The first choice went to the winner of the flip of a chip with spit on one side. A green bottle of corn whiskey usually circulated freely as opposing groups readied for the contest. (John Frierson, a planter near Mayesville, South Carolina, never allowed any drinks stronger than coffee at his shucking bees.) Sometimes whiskey and cider jugs were stashed at the bottoms of

THE SHUCKING depicts a husking bee, one of the most popular and universal pioneer social occasions.
CENTURY MAGAZINE, October, 1882.

corn piles for celebration at the finish. At times too, the host was carried around the heap of corn by two strapping men while the entire party snaked into line and sang "I'm gwin' to the shuckin' of the corn." Slaves sang their own folk songs, such as:

> All dem putty gals will be dar,
> Shuck dat corn before you eat,
> Dey will fix it for us rare,
> Shuck dat corn before you eat,
> I know dat supper will be big,
> Shuck dat corn before you eat,
> I think I smell a fine roast pig,
> Shuck dat corn before you eat. . . .
> I hope dey'll have some whiskey dar,
> Shuck dat corn before you eat,
> I think I'll fill my pockets full,
> Shuck dat corn before you eat.

Many other songs, often "rude" and "dirty" ones, were sung at the shucking bees. Such farm work songs constituted a genre in themselves, somewhat akin to sea chanties in their bawdy content. John Greenleaf Whittier's "Corn Song," from his longer poem, "The Huskers," was probably sung at these frolics. Team members often chose captains for their ability to lead in singing.

With a given signal, the rival teams would fall to work, furiously tearing loose husks with the aid of "shucking pegs" (an old Indian invention), snapping off the fibrous sheaths at the bases of the ears and throwing the grain-studded ears into baskets, crib, bin, or granary. Often the contestants engaged in cheating or horseplay—tossing in unhusked ears in order to save time; "misdirecting" throws, which chanced to hit opponents; heaving or kicking unshucked corn into the pile of the opposition. Challenges there were, and sometimes fights over these tactics. A red ear of grain (called a pokeberry ear) was a free ticket for a kiss posted on the cheek of the lucky finder's favorite lass or lad or for the choice of a dance partner. The same red ear might be surreptitiously recycled a number of times. The din of shouted encouragement and banter, thumping ears, and perhaps banjos and fiddles echoed through the encircling darkness. Upon finishing this part of their labor and merriment and awarding the prizes, the group might form a circle for a few rounds of "who's got the thimble." Or more probably, they would adjourn to long tables and gorge themselves on food prepared by the women. Hog and hominy and squirrel pot pie were among the favorite menu items.

After eating, the frolickers would clear the floor and scatter corn-meal over it for a barn dance. They would partake liberally of potent liquid refreshment and dance until the late hours, when they would file away in the gloom toward home, two by two, "as in Noah's Ark."

In newly opened frontier areas where women were scarce, "the frolic ends with a *stag dance*; that is, men and boys, without females, dance like mad devils, but in good humor, to the tune of a neighbor's cat gut and horse-hair." There are scores of descriptions of shucking bees tucked away in the obscure writings of pioneer America. And not all writers took kindly to these frolics. "If you love fun and frolic, and waste and slovenliness more than economy and profit, then give a husking," wrote a New England farmer. "Sing dirty songs for the entertainment of the boys," and expect your corn to be "mixed, crumbled, and dirty; some husked, some half husked, and some not at all." Typical of a number of frontier activities, the shucking bee combined work, play, food, drink, and a chance to mingle with the opposite sex.

In faraway Switzerland, long isolated from the beloved corn culture of his homeland, poet-diplomat Joel Barlow composed a lengthy poem entitled "The Hasty Pudding." In lines from this rhythmic narrative of 1793 Barlow presented his version of the husking bee.

> The days grow short; but though the fallen sun
> To the glad swain proclaims his day's work done;
> Night's pleasant shades his various task prolong,
> And yield new subjects to my various song.
> For now, the corn-house fill'd, the harvest home,
> The invited neighbors to the *husking* come;
> A frolic scene, where work, and mirth, and play,
> Unite their charms to chase the hours away.
> Where the huge heap lies center'd in the hall,
> The lamp suspended from the cheerful wall,
> Brown, corn-fed nymphs, and strong, hard-handed beaux,
> Alternate ranged, extend in circling rows,
> Assume their seats, the solid mass attack;
> The dry husks rustle, and the corn-cobs crack;
> The song, the laugh, alternate notes resound,
> And the sweet cider trips in silence round.
> The laws of husking every wight can tell,
> And sure no laws he ever keeps so well:
> For each red ear a general kiss he gains,
> With each smut ear he smuts the luckless swains;
> But when to some sweet maid a prize is cast,
> Red as her lips and taper as her waist,
> She walks the round and culls one favor'd beau,

Who leaps the luscious tribute to bestow.
Various the sports, as are the wits and brains
Of well-pleased lasses and contending swains;
Till the vast mound of corn is swept away,
And he that gets the last ear wins the day.
Meanwhile, the housewife urges all her care,
The well-earn'd feast to hasten and prepare.

It happened on a Sunday afternoon at one of those grizzly, un-painted barns, its lichen-spotted shingles and curled boards having the look of some prehistoric structure grown from the earth at the dawn of time. Neighborhood youngsters, worn from a week's hard work and released from worship, chose sides and squared off for a leg-endary corncob fight. One team drew the loft as home base, the other the ground floor and adjoining barnyard. Knotholes became peep-holes; doors were portholes, grindstones, toolboxes, animal stall parti-tions and water troughs were barricades. The combatants, carrying salt bags, "gunny" sacks, or pocketfuls of cobs, stalked each other for surprise attacks.

Except perhaps for the snowball and the spitwad, no projectile in all the world was so ideally suited for doing battle and relieving aggres-sions without injury to the parties involved as the corncob. True, it was too long and unwieldy and light to throw accurately. But some rivals would break each cob in half and dip the pieces in water for greater weight and accuracy. One could hear the near-misses "whoosh" past and at times feel the spray of water. After an hour or so of warfare, someone would shout "Let's switch places," and a truce was observed until upstairs and downstairs crews had reversed posi-tions. There was no scoring in the game—only personal triumphs from hits or near-misses and release of pent-up energies.

Sometimes grown men engaged in such battles. William Clark "Charles" Quantrill and his band, between their savage guerrilla raids during the Civil War, were known to sharpen their bushwhacker in-stincts by staging corncob fights. It happened on a thousand Sunday afternoons, at a million old gray barns, this jousting of gladiators in an almost-forgotten sport of a bygone era. Barns are not built that way anymore, and corncobs, for the most part, are chopped fine by ma-chinery for a host of modern uses. Unmentioned in modern historical accounts, the corncob fight was a significant phenomenon in the so-cial structure of pioneer America.

Nor were cob fights the only recreational uses of these pithy cores.

The corncob fight was an important social outlet in the predominantly rural society of pioneer America.

Corncobs were the universal frontier back scratchers, as necessary to farmers as were fenceposts to hogs. Light in weight and with soft centers which could be easily punctured for lines, cobs made good fishing floats or bobbers. Below these cob corks, fishhooks were often baited with grains of soft, milky corn, particularly when anglers were after catfish. Every rural youngster knew the wonderment of throwing cobs at leather-winged bats in the gathering twilight and watching these uncanny creatures erratically dip and swirl and track the hurled objects as if they were tasty insects. Always the bats seemed to veer away just before the point of contact. Kids would slide down a pile of corn-cobs with all the glee of tobogganers on a snowy slope, or otters play-ing follow-the-leader on a mudslide. In some areas this was as popular as jumping into the hay mow.

Then there were the hay rides—groups of young people bunched together, always in mixed company, atop cushiony wagonloads of hay, with horses or oxen pulling at a slow gait under romantic moonlit skies. They were always called hay rides even though the joy riders often sat on high-piled bundles of dried corn fodder. The stalks in their vertical posture were also objects of no little recreational interest. Tall, luxuriant corn plants stood like a dense forest just a stone's throw beyond the cabin door and were ideal for games of follow-the-leader and hide-and-seek or for concealing the trysts of lovers who did not wish to be chaperoned. The young getaways had only to be cautious about visibility down the checkerboard corn rows. Barlow wrote as if he had been among the long, green aisles.

A corncob fishing float and a hook baited with green corn for catfish.

High as a hop field waves the silent grove,
A safe retreat for little thefts of love,
When the pledged roasting ears invite the maid,
To meet her swain beneath the new-form'd shade.

Wigwam-shaped corn shocks offered a whole new dimension of fun magic. Every indulgent frontier parent would allow each youngster in the family a shock to remodel as a house, a private, cozy nook shared only by rats, mice, and a few other wild creatures, and perhaps by a cat or dog. A child of not-too-bulky proportions could crawl into a shock above the waistband of vine or twine, close the opening with corn leaves, and settle into a warm, comfortable, secluded hideaway for daydreaming or hide-and-seek. Merrymakers have used shocks and stalks for decorations at barn dances and other rural festivities throughout American history. Both Hernando de Soto and John Smith observed on their early travels in America that Indians sucked and chewed on sweet green corn stalks as a kind of candy, just as sugar cane was used farther to the south. Many a frontier youngster made a cornstalk fiddle by slipping a wood bridge under loosened fibers of the stalk.

As with cob and stalk, the fruit of the corn plant, its golden grain, has been widely used for fun and games, from grains of candy simulating corn, still abundant at Hallowe'en time, to giving prizes for the person who guesses closest to the number of grains in a jar. Children played going to the mill by trying to guess which of the opponent's closed hands held the taw, or grain of corn. Each right guess equaled one mile. And they conjured up games of throwing corn, like hull-gull or hully-gully. Strings of popcorn were used as decorative objects thousands of years before Columbus, and white and black Americans followed the practice, particularly in making festooned ornaments for Christmas trees. Amateur jewelers fashioned grains of various colors into necklaces, bracelets, and pins (still available to tourists in the Southwest). A favorite trick for teaching stray dogs and cats and even human pranksters to beware was to load a shotgun with small, round popcorn grains. When fired at fifteen to twenty yards range, these light pellets would sting without penetrating dangerously. Combined with a deafening surprise explosion, they usually had the desired effect. Frontier anglers and hunters used corn grains as bait for fishing, hunting, and turkey trapping. Beano, or bingo, was a numbers game most commonly played with corn seed or beans, and here and there at county and state fairs the cries of the game caller and players can still be heard: "I-29; B-7; bingo!" "Don't move the corn. There may be a mistake!"

County fairs and corn displays were as closely identified with each other as grains and cobs. Elkanah Watson of Massachusetts introduced the agricultural society concept in 1811. It spread rapidly, and by the 1830s societies were giving awards at fairs for the largest, fullest, and most even ears of grain. And there were prizes for shucking and for cutting an ear cleanly from the stalk with a throw of a corn knife. After the Civil War, journalist and political figure Horace Greeley urged agricultural societies to give prizes for the best farm products. Late in the nineteenth century the economics and socialization of corn would be further wedded by Seaman A. Knapp's 4-H clubs and "Professor" Holden's "Seed Corn Gospel Trains," which were nothing more than corn fairs on wheels. Boosters from Sioux City, Iowa, in 1889, chartered a train, outfitted it as a rolling corn exhibit, and crossed the nation to take part in President Benjamin Harrison's inauguration. Corn palaces, frame buildings completely covered with ears of husked corn, have been built in a number of corn-belt states. Among these was the palace at Sioux City, which was visited by President Grover Cleveland in 1887. Another such structure stands at Mitchell, South Dakota. These grain-covered edifices have proven attractive to hordes of tourists and flocks of crows, blackbirds, starlings, sparrows, pigeons, and other winged invaders. Because of these latter, corn palace carpenters have occasionally found it necessary to reshingle parts of the buildings with fresh ears.

There have been other pleasures derived from America's corn crops. Corn-husk dolls were favorite creations in western homes, and these, when they have survived the forces of time, have become collectors' items. The pleasures corn has given in eating, drinking, smoking, literature, poetry, and music, discussed elsewhere in this volume, were also unquestionably large. Indian corn and human life in America have been so thoroughly interdependent that the crop almost literally filled the human experience. Even in retrospect, it is not possible to discern where work ended and fun began, for there was fun in work and work in pleasure. No doubt the pioneers considered eating and drinking the processed corn products as the ultimate in corn-related pleasures. Yet one cannot escape the conclusion that is conveyed between the lines of early settlers' observations on the subject that without the toil those pleasures would have been materially lessened. A frontier Missouri corn farmer of the early 1800s expressed his doubt "that pleasure could be perceived without its contrast. . . . They must alternate before we have a consciousness of either." Given the many work stages in corn raising, there was no dearth of labor consciousness. Each stage, from clearing the densely wooded land, to harvesting and final processing of the crop, was replete with its drudgery, its revelry, and its ritual.

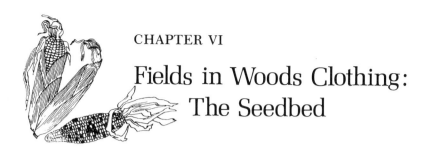

CHAPTER VI

Fields in Woods Clothing: The Seedbed

The Whole Country is a perfect Forest.

HUGH JONES in the Ohio
Valley, 1724

To the newcomer, fresh from Europe's populous civilization and its fields cultivated for eons, the most striking feature of America was the awesome forest land. True, there were coastal belts and inland patches that had been cleared for farming by Indians, and hundreds of miles into the hinterland there were oak openings, prairies, and plains as awesome in their own way. But overall, from the Atlantic to the Great Plains, the forest owned the land. The scene for the first two hundred years after settlement at Jamestown was one of endless stands of timber—oak and hickory, walnut and maple, beech, birch and ash, elm, sycamore, pine and cedar, and scores of other species. Their seasonal variations were a mottled canopy of leaf colors. The pioneers looked upon those vast sylvan landscapes with mixed feelings. They sheltered wild game so essential to the settlers' survival, and their trunks and branches served as cheap materials for cabins, barns, barrels and buckets, furniture, fences, and fires. Yet they shielded Indian, French, and Spanish adversaries who from time to time challenged the westward march of the settlers. Finally, and all too quickly, it was necessary to convert forests to fields.

For the first corn-farming colonists some lands had already been cleared. In Virginia, Massachusetts, Maryland, and elsewhere, English immigrants took advantage of abandoned Indian cornfields. Faced with a shortage of iron implements in the first years, they relied in part on the natives' tools to loosen the soil for planting—sharpened sticks and wood-handled hoes of flint, bone, shell, and stag horn. The technique was primitive, chopping weeds and sprouts, loosening dirt and forming it into small "hills" where seed grain would be planted. These were emergency measures of the first arrivals who were clinging to life by a thread. With increases in their numbers, the early colonists soon faced one of the major tasks that would confront them and their successors for the next two centuries and more, that of clearing away the virgin forests. They hacked out sapplings; they cut giant monarchs that may have shaded LaSalle and Marquette, Lewis and Clark, and a hundred Indian tribes before them. And they were so successful that, as well as clearing millions of acres of farm land, they created the myth that only timbered soil would grow crops.

Several basic unwritten rules of coping with the primitive environment emerged from the colonial experience, among which were: farm the forest lands first because they were believed to be the most fertile and, whatever the long-range plans, start with corn since it was staple for both people and livestock and was the most adaptable grain crop. (Wheat did not thrive in newly cleared forest land.)

Often the family patriarch, perhaps accompanied by a son or a slave or several other heads of families, would go on foot, on horseback, or by canoe to scout and lay "corn claim" to a new homesite. Having mapped a route and located land for his dream cabin and, if time permitted, cleared some of the land, he would return for the family. It was as inevitable for land-starved settlers to probe into the wall of virgin forest as it was for hogs to root under loose fence rails. Customarily several families, often relatives, made the permanent move together for mutual protection. Perhaps their way west was along some ancient animal trail converted to an Indian path, then rutted by immigrants with white-sheeted wagons. Such were the beginnings of Kittanning Path or Forbes Road in Pennsylvania, Braddock's Road leading northwest from Virginia, or Cumberland Gap and Wilderness Road from North Carolina and Virginia to Kentucky and Tennessee. Or they may have traveled down the Ohio, Tennessee, or Kentucky River in a big, boxy, homemade flatboat. This craft was a one-way disposable container that offered some watery protection from Indian attack and could be disassembled and used for buildings at the homesite.

Despite the cold weather, late winter or early spring was generally the best time to leave, as dictated by the corn crop. It was important for the family to do the traveling between harvest time at the old farm and planting at the new. The experiences which resulted were sometimes tragic. A black slave of Thomas Hutchings' froze to death on March 6, 1780, on a Tennessee River flatboat float to Cumberland corn country. But most migrants considered it better to risk a few weeks of freezes than a whole season of famine without corn. Winter travel also reduced the likelihood of ambush by Indians or outlaws, for the leafless trees and bushes provided them no cover.

Arriving at their wilderness site, the weary travelers had not a moment to rest on their achievement. Two almost simultaneous needs confronted them—defense, and getting their corn in. They would quickly rough out an individual family cabin or a "station" or stockade, leaving interior refinements for less pressing months. In times and places of slight danger, the newly arrived settlers lived in (and out of) wagons, flatboats, or tents and turned quickly to the corn crop.

Clearing a large, heavily timbered area of perhaps five to twenty acres for a corn crop in a few days sounds practical for a modern-day gang of laborers with chain saws, dynamite, and bulldozers. The task was more difficult for the new settlers to accomplish with simple tools, under danger from Indian attack, in time for a spring planting, and with no equipment heavier than an axe and grubbing hoe. Early colonists, having had no European experience at clearing forests for immediate planting, turned to their then-friendly Indian neighbors for techniques. Within a short time the practices were standard throughout wooded America. The smaller trees were easily and quickly cut for building. Weeds, grass, and underbrush had been kept in check by the canopy of leaves from the trees that screened out sunlight. But the giants of the forest defied rapid clearing with axe and saw. The farmers simply girdled them. An eight- or ten-inch strip of bark was either "bruised and burned" or chopped off completely around every tree, cutting the life-giving arteries. It was such a quick operation that a family could kill several acres of forest in a single day, and thus save much labor when time was very precious. Frequently a group of families girdled trees cooperatively, in what pioneers referred to as a frolic.

An area of these bare, standing skeletons was called a "deadenin'" and the land was spoken of as newground. The Indians had another way of quickly reducing a forest to a deadening, which the pioneers adopted, often in conjunction with girdling. Using grubbing hoes, axes, mattocks, cane knives, and corn knives, they dug and cut brush, cane, the ever-present dead branches, and slash trimmed from cabin-building logs. Along with dead leaves, they piled this refuse at the bases of large trees, and when it was dry enough, preferably on a windy day, set it afire. Burning reduced the brush piles to ashes and killed the trees at the same time. Crusts of charcoal were chopped off and the trunks burned in the same manner again and again until the trees toppled.

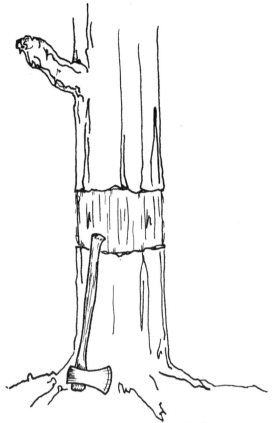

Trees were killed by girdling, and the first-year corn crop was planted in the resulting "deadenin'."

Swedish settlers along the Delaware River brought with them a long-used Scandinavian practice of timber clearing. It was called *svedjebruket*, which simply meant burning. They would cut the trees and let them dry for a year. After removing useful logs, they would pile and burn the tops and branches. Of course this method cost them a year's crop as compared with girdling and deadening.

Some farmers had enough foresight to plan their fence lines when clearing timber for the plow. They cut trees of proper spacing, size, and species four or five feet above the ground. The tall stumps were excellent fence posts.

As if the clearing of a large forest of big hardwood trees and underbrush were not a complex enough task by itself, there was a further problem. Throughout the corn country the woods harbored a mixed blessing, tangled growths of wild grapevines. There was a reason the Norsemen referred to this new world as Vineland! Except for their tart fruit, which put a spirited flavor into pioneer menus, the tie cords which they provided for corn shocks, and the delightful swings that children relished, grapevines were cursed because of their resilient lacework, which wove the forest together. They would run along the ground for fifty feet, then climb the tall trunks, and reach from one tree to another. It was a common occurrence for woodsmen to cut a tree, then shove it in dismay as it refused to fall because of the grapevine ties to its neighbor trees. But again, the deadening was the answer. By the time the wooded giants were dead, so were those clinging vines.

The soil in a first year deadening or newground, blanketed with the moss and leaf mold of yesteryear, was loose enough to raise a fair corn crop without plowing. This was important since there were other demands for labor at a new cabin site. All that was necessary was for the laborers to break the leafy, humus-laden earth at the points of seeding with a spade, hoe, or stick. Corn growth would not be seriously challenged by weeds during the first season since there was little residue of weed seeds, and the dead limbs admitted ample sunlight for corn except within several feet of large trunks on the north side. Corn thrived in a deadening better than any other field crop. The job was quickly and effectively done. John Muir reported that such a small clearing, intensively cared for, would produce enough corn to carry the family and its livestock through the first winter. Having staved off starvation, the new settlers could use their slack hours and seasons to girdle and grub a new deadening for the next year and work on clearing the old one. They continued the clearing of new ground annually until all the area ultimately intended for farming land had its trees deadened or removed.

A deadening was not a healthy place to be after a year or two. It was a battalion of standing skeletons, a boneyard of boles and branches. Practically everyone knew of someone who had been hit by a falling dead branch. "Stay out of a deadenin' in a thunderstorm," and "never tether your horse in a deadenin'" were religiously obeyed frontier rules. Small children were warned against playing in these graveyards of trees. The Seneca Indians had a special prayer for the safety of their children in deadenings.

Eventually, big as the job was, a deadening had to be cleared. Many limbs were pulled off before they fell of their own weight. Workers used crooks and wooden and metal pruning hooks with long handles for this purpose. The old expression, "by hook or by crook," probably came from the use of these pruning tools. For breaking a limb beyond the reach of hook or crook, settlers used an effective device, a rock tied to a rope. It was a simple matter to throw the stone over a branch, then work it down within reach by flipping the rope, holding both ends of the line and pulling.

Thus did the advancing farmers rob the timber of its long-held title to the land. But even in death, the heavy trunks and limbs resisted. Like the bottom of Thoreau's Walden Pond, the earth was "strewed with the wrecks of the forest." Biting axe, rasping saw, and charring flames gradually disposed of the wreckage. Dead trees were cut and used for building, fencing, and firewood. They made much better fuel for the fireplace than green cuttings. Much of the residue from this clearing process was piled and burned in the field. Clearing by fire, more characteristic of Indians than settlers, was hard on the soil, and fields prepared in that way were usually abandoned after a few years for lack of productivity. This may have been partly responsible for abandonment of the milpas, the cornfields of the Maya Indians in Central America.

Ideally, a deadening would be ready for the plow by the second or third year when soils had become so packed that they needed deep loosening. This was no small order. Forests relentlessly strove to re-store themselves, and while girdling or burning killed the trees, the roots remained alive for several years. A ring of shoots, called sprouts by the pioneers, grew around each stump or trunk below the girdle cut. In just one season these would become small saplings, and there would be ten or twenty on each stump. If unattended, a rank jungle would soon replace both forest and stalks of corn. Fortunately for the frontier farmer, these sprouts were not difficult to deal with if not al-lowed to get too far out of hand. One could kick them off easily at the stumps or knock them off with a stick or hoe. The process had to be repeated many times before the roots died. This task almost univer-sally fell to the children, who of course were sternly warned about fall-ing limbs.

Some stumps were so large and heavily rooted that farmers plowed around them for years before rotting finally took care of the problem. Oak, hickory, walnut, and hard maple would remain for thirty to fifty years, and stump pullers using leverage and horses or oxen had to be devised. Large roots near the surface were grubbed free and chopped with felling axes to ease the burden of extracting. In the late 1800s dynamite took over some of the stump pulling chores. Roots protruded in such jagged patterns from pulled stumps that farmers were known to haul them to the borders of the fields and line them up as fences. The complete clearing of a forest was so prolonged that often as not the soil became tired and the restless farm family moved on before it was done.

As if the trees were not obstacle enough, the tillers in some areas were confronted with rock-infested soils. Known to geologists as *drift* from continental glaciation of the Pleistocene age, these rocks were carried southward as glacial sheets advanced, and then settled on the soil when the ice melted. They extended throughout New England and New York, into northern New Jersey and northeastern Pennsylvania, and over much of Michigan, Wisconsin, and Minnesota. Glaciers deposited smaller numbers of stones in the states bordering the Great Lakes on the south and as far west as Iowa. Alluvial fans and aprons at the bases of all the mountain ranges presented some rock problems of a different type. The washed-out rocks in such areas were smaller than many of the ice-deposited boulders. Northern Europeans had some acquaintance with glacial and piedmont terrain. As with timber, however, the Old World fields had been long cleared of rock, and those of the New World thus posed unaccustomed problems.

Pioneers found rocks harder to clear than timber, since they would not burn or rot. If they were too large to move, they could not be cut down to smaller size, and there were always more than met the eye. The wearied toilers estimated that a typically glaciated acre of New England soil required the labor of one able-bodied adult for sixty days to clear the surface rocks. After one season's cultivation and a winter of freezing and thawing, enough additional rocks had worked to the surface to require another sixty days for a laborer to remove. Even that was not the end, as more stones would work their way upward for a number of seasons to come.

Something of the tenacity of the sturdy Puritan pioneers and their commitment to work can be appreciated when one considers the size of their task. They hauled small rocks on handbarrows, wheelbarrows, or wagons. Stone "boats," actually low sleds, were better than wagons for moving heavy boulders, since workers could more easily pry and roll the rocks onto the low boat decks than lift them thirty inches or more to a wagon bed. The sleds were pulled by draft animals over dirt, mud, or snow to be dumped in low points or gullies. Thousands of New England and New York farmers sledded stones to the borders of fields and there fitted them carefully together into stone walls or fences and topped them with wooden rails for greater height. Remnants of many such fences have survived to the present day. They represented as much a desirable way to dispose of stones as a need for fences. It was essential to remove the rocks in order to protect the cutting edges of spades, plows, and other tools used to work the soil.

Implements for breaking ground have been in use for thousands of years in many parts of the world, but nowhere were they more essential than in preparing the seedbed for corn. While this giant, exotic grass produced far better than any other grain in the new ground of a deadening, it was a fast-growing crop with roots extending three to four feet and more. In packed soils it required deeper spading or plowing than wheat, rye, barley, or oats in order to permit better penetration of the earth by roots, moisture, and air. Turning the soil had the additional advantage of folding the weeds under, killing them, and incorporating this humus as a "green manure" or natural fertilizer. By the second and third years after killing the trees, this was important since weeds had had time to infest the new ground.

Except in Peru, where crude spades had been developed, Indians of the Americas were not able to break and turn the ground as the Europeans did. This helps to explain why their corn culture succeeded best in the naturally loose, sandy soils and the newly cleared loams, which had not become packed from exposure and cropping. It also accounts in part for their abandonment of fields which the colonists were able to farm successfully and for the fact that the Indian farmers did not grow much corn or any other crops in the heavy prairie and Great Plains sods.

Spading was the simplest and most thorough land-breaking technique. Settlers used metal, and even oak- and hickory-bladed spades in all areas of corn culture. As early as 1622 spades were listed among necessary items for English colonists coming to America. With spades farmers could control the depth of digging and work around obstacles such as stumps, roots, and stones, carefully turning weeds and grass under. Back-sodding, a process similar to spading but done with a heavy spading hoe, was employed particularly by slaves in some corn areas of the South. Rebecca Burlend, English immigrant to Illinois in the 1830s, relates that the family's mare was foaling and could not pull a plow. In desperation, the farm woman and her husband and ten-year-old son back-sodded four acres and planted it to corn with no other implements than hoes. Garden-size plots and fence corners could be handled by spading, but except in emergencies it was too slow for large fields.

When the mold of the earth was right for plowing there was no time to waste, and for all its limitations and imperfections, it was the plow which had to cope with rock- and root-infested fields. It was never as simple as harnessing the team and turning furrows at a given date on the calendar. Writing in 1790, Samuel Deane of Worcester, Massachusetts, advised double plowing, fall and spring, as the best preparation for corn planting. As for spring plowing, sometime after sassafrass-root-digging and the end of trapping season the settler could be seen walking about the fields, stooping, picking up handfuls of dirt, rubbing and crumbling it between fingers and thumb. Samples of "mold" at high and low points of land were tested. Too much crumble, bordering on dust, and the plow would have to wait for rain; too sticky, and the mold board would cause the soil to "bake," clodding and ruining the land for perhaps several years to come until freezing and thawing broke the clods. Better a trifle dry than too wet. Ideally, the earth would be dark, moist but crumbly at the same time the grass had grown just high enough that it would not turn and resprout when plowed under.

It was always a chancy business. A wet spring meant late plowing and planting and danger of damage to the crop from early fall frosts. No help from a radio or television set or daily newspaper weather report; just a few words on "weather sign" from the farm almanac or settler's guide, for those who were lucky enough to have them and to have a family member or neighbor able to read. Experience and legend were the most trusted guides. And so the anxious farmer kept an eye on the soil and a "weather eye" on the sky. "Any fool" could recognize rolling thunderstorms and solid banks of rain clouds. But the farmer needed more warning than wet buckskin or soaked homespun clothes. Real or imagined, there were signs the settlers swore by. Red sunset or a few clouds obscuring the sun in the east or a heavy night dew foretold clear weather. Among the countless warnings of rain were a rainbow in the morning, yellow sunset, a shift of wind to the east, a circle or halo around the moon or sun, low-hanging smoke, low-flying swallows, intensified smells of cellars or swamps, excitement and restlessness among birds and animals, ants covering their tunnels, earthworms coming up to the surface, and high, filmy (cirrus) clouds followed by mackerel sky. Modern tractors pulling gangs of implements over land which can be irrigated or rain-bird-sprinkled to suit the need have taken much of the chance and guesswork out of farming, but the early settler was a prisoner of the elements. Like a feeding rodent watching for a bird of prey, the farmer's life depended largely on the ability to "read sign" both below and above.

Having judged that the best combination of conditions prevailed, the pioneer would literally leap into action. Oxen were yoked, horses were bridled and fitted with collars (padded with dried corn husks), traces were hooked to whippletrees (singletrees) and these to the doubletree, and clevis and clevis pins linked the animals to the plow.

A crude plow it was on the typical American farm before the 1840s—much like those of ancient Egypt or Mesopotamia or China. Often it was simply a sturdy oak or other hardwood fork, with one end eight or ten feet long for the beam and the other prong short and sharp for the share and moldboard. Handles were attached at the fork. This device merely dug in and loosened a series of closely paralleled trenches across the field; it did not turn the weeds under. Rare was the farmer or planter who could import one of the heavy and more efficient wood and metal plows of medieval and early modern European origin. These were expensive and Americans relied largely on copies and designs of their own. Heavily bound as farmers have been to tradition, they changed plow designs slowly and reluctantly, calling upon superstitions to support their uneasiness about anything new. In colonial times the typical plow had a ten-foot hardwood beam and a wood moldboard (the curved piece which turned the furrow over). This board was cut and hewn from a winding tree which had the proper curve. Like the Old Colony Strong Plow of the 1730s, it was often faced with old hoe blades, worn horseshoes, or other metal pieces which would increase the wear qualities. A steel point or share, which cut the earth to be turned (or rather "stirred") was fastened to the bottom front of the moldboard. Sharpening the share was usually a blacksmith's task. Many plows had vertical handles, which made them difficult to manage. Converted from pounds sterling, the cost of the machine was about the equivalent of forty to fifty dollars, a major expense considering the specie shortage. Furrows which at best were poorly turned by this implement, protruded like "ribs of a lean horse" according to one account.

In a country where necessity was of epidemic proportions, new plow designs flooded the patent office. Charles Newbold's device of 1796 was solid cast iron and a vast improvement over the typical colonial model, but despite his expensive promotional campaign ($30,000), farmers scorned it because of a totally unfounded superstition that the iron moldboard "poisoned" the land and stimulated weed growth. In 1817,

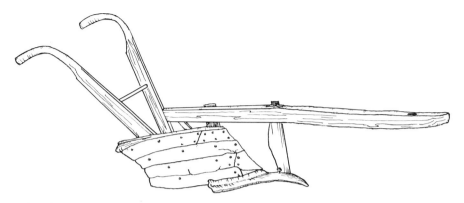

*The Old Colony Strong Plow, made in New Hampshire in 1732, was one of
the type that turned furrows in fields previously cleared of trees and rocks.*

Jethro Wood did for plows what Eli Whitney had done with gun manufacturing, introducing mass production by interchangeable parts.
Wood's plow was Wood in name only: it had cast-iron moldboard, landside, and share. Because the cost could be kept down to only six to eight
dollars, it changed corn culture greatly. Western centers such as Columbus, Ohio, sold it by 1825. A number of other designs soon hit the
market and plow names such as Diamond, Carey, Tobey and Anderson,
Clark, Jewett, and Daniel Webster achieved some recognition, as did
the Andrus steel plow of 1837.

Still the former standbys, in particular the jumping-shovel plow and
the bull or bar-share plow remained in very wide use in the West and
Southwest. Southerners and many farmers of the West preferred the
jumping-shovel plow. Its curved coulter, a cutter riding ahead of the
shovel, sliced through small roots. When it struck roots too large to
cut, the coulter caused the share to jump or ride over the roots and
resume its normal depth of cut beyond them. B.C. Wailes of Mississippi noted in 1826, "this plough will break up thoroughly and with
ease, the almost impervious mat of cane roots. . . . It is particularly
serviceable in planting the first crop in cane land." The jumping plow
was most effective in a new deadening, where absence of grass and
weeds made it unnecessary to turn the furrow upside down. "Throwing the dirt on both sides," it loosened the soil in a series of closely
parallel trenches and ridges.

New Englanders preferred a heavy plow with six or seven yoke of
oxen which, with a man seated on the beam to hold it down, reportedly could cut through a four-inch oak root. When the share or the

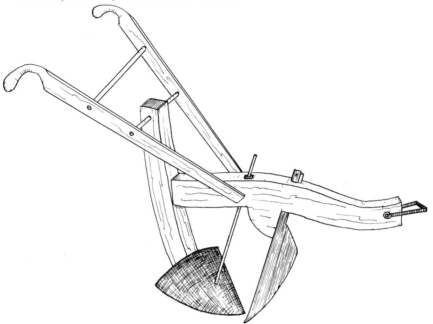

The jumping shovel plow was remarkably well adapted to loosening root-infested soil in deadenings and new clearings. The coulter blade in front of the shovel, which could cut roots up to three to four inches in diameter, jumped over larger roots.

shovel of any of these implements struck a massive root or rock, the plow operator was often thrown violently against the bar between the handles. Frequently, roots would stretch and snap, striking the plowhand in the shins, and black-and-blue legs were a standard part of the chore. At times, coulters and share points stuck in large roots and required the tuggings of several workers to dislodge. Axe wielders often accompanied plowhands to help cut the roots. While men did most of the heavy work, women and children on occasion took their places at the plow handles, and sometimes the efforts of four or five laborers were required at once. As one writer expressed it, the customary language of animal control, "giddap," "whoa," "gee," and "haw," were supplemented by some colorful expressions not normally quoted. A profane plowman commented to a passerby, "That——mule just don't understand no other language!" An old farm saying held that God took care not to notice the language of mule drivers.

The heavy sods of prairies from northern Indiana westward and on the Great Plains presented a new set of problems. Settlers often made their first plantings by chopping holes in the sod with axes. There was little other use for these tools because there were few trees to cut. Prairie farmers believed that the deep corn roots helped to loosen new sods for easier plowing in succeeding years. Sticky prairie sods called for sturdy steel plows. Most effective for initial breaking were implements that could cut and overturn furrows perhaps eighteen inches wide and only four inches deep. After the roots had decomposed, these soils could be plowed deeper, but the rotting humus gummed up moldboards, which had to be scoured by hand frequently, else they would not turn the furrows properly.

In the heavy sod areas of the prairie, a professional group of plowhands known as "breakers" emerged. They moved from one location to another, plowing the sod for $2.50 to $3.00 per acre. They were usually young, unmarried men who were earning money to buy their own farms. The nature of their equipment is a rough index to the difficulty of breaking sod: three to seven yoke of oxen; a plow with a heavy, fourteen-foot wood beam and a 60- to 125-pound iron or steel plowshare. The furrow turned by this rig varied from sixteen to thirty inches wide and two to six inches in depth.

Scouring was a key problem in plowing. A smooth scour was essential to a well turned furrow that put weeds and grass deep underneath. John Deere, a blacksmith of Grand Detour, Illinois, largely solved this problem in 1837 by making moldboards out of German mill-saw blades. It was a number of years before he produced these plows in commercial quantities. After mid-century, various improvements

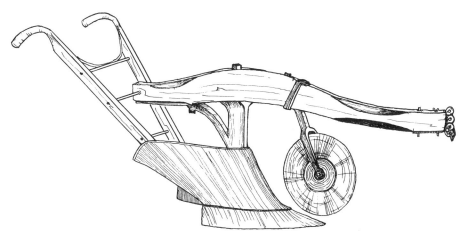

The John Deere Company's Moline plow was made in the mid-nineteenth century. With its sharp rolling coulter, it cut and turned smooth, neat furrows.

brought the plow to a modern stage of efficiency—a sharp, rolling coulter that cut a clean furrow on the landside and a polished, full-curved moldboard for a perfect turnover of each furrow, as exemplified by the Moline plow; a loop chain, which pulled all weeds completely underneath the dirt; and a curved steel I beam with a row of clevis pin holes for controlling depth and width of furrows. Among the other late nineteenth century innovations in plow design were the two-wheeled sulky plow (1864); the walking gang plow (1867); a moldboard with soft center and hard, heat-treated surface (1868); the chilled-steel plow (1870); and the foot-lift riding plow (1899). The double bladed lister plow (1880s) turned two furrows toward each other forming a ridge, at the top of which the seeder planted a row of corn. Later, when time permitted, the unplowed section between these ridges could be broken up with a mule and a "middle buster."

A few corn yield statistics demonstrate the importance of the plow. Nonplowed newground produced up to 30 bushels of corn per acre the first year, a worthwhile bargain under the work pressures at a new homesite. Second-year plowing upped the yield to as much as 50 bushels. This was doubled to 104 bushels to the acre by Christopher Leaming of Ohio who resorted to deep plowing (two or three inches below normal depth). And while sodbusters hard pressed for time could produce 10 to 20 bushels to the acre by chopping planting holes in prairie sod with an axe, it was obvious that catering to corn's deep roots paid off handsomely.

Farmers have a reputation for being individualistic and conservative in their habits. All through the pioneer period this resulted in widely varying practices in most aspects of farming. Writers almost invariably recommended deep plowing for corn. "Plague on shallow plowing. . . . I have ever found deep plowing attended with good crops," wrote Weatherwise (the *nom de plume* of Robert Thomas) of New England. The advice of Horace Greeley, Edward Enfield, and all the farm journals echoed those sentiments. Yet settlers clung tenaciously to their own ideas. Often they did what was easiest and cheapest in the short run, even to the extent of adopting methods enabling them to avoid turning the soil of the entire field's area.

Some farmers tried a compromise between hill-digging and plowing an entire field, resorting to crosshatch plowing, which left four- to nine-foot squares unturned. They ran a shallow furrow down each plowed ridge and planted corn at the intersections. The unplowed interiors of the squares could be left or plowed later. This was the most common method in the colonial South and North alike, although Pehr Kalm observed about 1750 that the best farmers plowed the entire field at one time.

Farmers would usually alternate from year to year in a given field, starting on the perimeter of the field, turning furrows outward, and continuing in ever decreasing circles until they reached the center of the field, then dragging dirt into the resulting low point at the center of the field with the plow on its side. The next year, the plowing would begin at mid-field, with furrows turned toward the center. But for this alternate moving of the soil back and forth the width of a furrow, eventually the field would have been either heaped into a mound at the center, or cut into a shallow bowl.

Every pioneer plowhand who walked miles in spring furrows, often barefooted, behind a team of mules or horses or yokes of oxen, experienced some surprises that rolled out directly underfoot in the fresh-cut trenches. Here a mole and there a field mouse, and occasionally a stunned, writhing six-foot "pilot" snake (black snake), a corn snake, or perhaps a lethal copperhead that had taken refuge in a mole tunnel. They usually dispatched serpents with the weed chain if the snakes were not killed by the share or coulter. Frontier farmers habitually slaughtered all except green snakes, king snakes, and the tiny, ring-necked or newground snakes.

Less pleasant, particularly for the team, was a nest of bumblebees, called humble-bees in early America. The animals would sometimes stampede after being stung, and each bee sting on horse or mule left a small patch of permanently white hair. The farmer had the option of leaving an area near the nest unplowed on succeeding rounds or destroying the bees. Hundreds of angry bees were not things to be dispatched with a fly swatter, but the settlers had a way. They drove the team a few rods distant, heated water, and poured a jug about half full. Placing the jug beside the nest with the aid of a long pole, they stirred the nest with the pole and retired to watch the sport. The angry swarm would buzz about, then as if drawn by an irresistible magnet, would stream with a gurgling noise down the neck of the jug to their doom. In a few minutes time the plowing could be continued.

Every plowhand, too, knew the thrill of turning up flint arrow and spear points, fleshing tools, and the occasional knife. More rare were Indian tomahawks, mortars, pestles, and even chunkey stones. Nearly

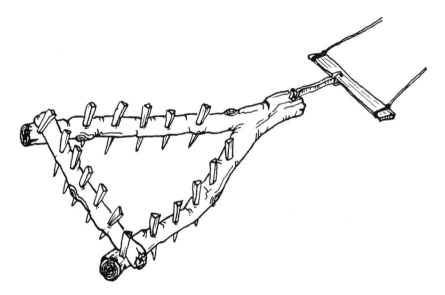

This primitive triangular harrow replaced simple tree-branch and log drags for breaking clods before planting. Later models, both triangular and rectangular, were made of iron.

every pioneer cabin had its pile of Indian relics turned up in the furrow, or perhaps exposed by the disking and harrowing of the plowed fields in preparation for planting.

Laborious as was deep plowing, the pioneer family had more to do before planting time. To cope with the furrow ridges, which were too rough and infested with clods, they fell back on ancient European experience, bringing into use the drag, or tooth harrow, and eventually the disk harrow. The simplest or harrows was a forked tree, the trunk of which was the tongue. Stubs of smaller limbs were left on the V-shaped branches. When dragged over the plowed field, these stubs broke the clods and smoothed the soil. More modern harrows were variations on the same theme: hardwood pegs driven through holes in triangular (A-frame) planks or platforms; tandem or double V-frames; iron harrow teeth in wood platforms; and eventually, spring-tooth harrows and all-iron frames with handles to adjust the angle of the clod-breaking teeth. Farm animals dragged these giant rakes across the fields.

Sometimes a driver would stand on the back beam of the harrow or lash on a log to give it additional weight. Double harrowing was usually done by a "half lap" across each track on the next round. Roots and grass clumps incessantly clogged the harrow teeth, and the teamster frequently cleared them by momentarily lifting the rig in motion. In newly broken fields this was "devilish hard" work. Eventually, farm equipment companies made wheeled harrow carts for farmers who preferred riding to walking behind the harrow, but as with the first plow seats, they were called lazy man's rigs.

The revolving disk harrow, invented in 1847, was designed to do much the same job as the tooth harrow. It consisted of beams that held rotating, dish-shaped steel disks, which chopped and smoothed the plowed furrows. On occasion, the soil of a field was soft and loose enough that the farmer would skip the plowing operation for a year and merely disk. This was particularly true when corn had been shocked or harvested, and winter wheat was planted between the shocks. Many farmers used both disk and harrow after plowing in order to pulverize the soil for better germination and growth of corn. Deep plowing in the spring required considerable working of the soil with harrow, disk, and sometimes with either roller or planker, not only to break clods but to recompact the soil sufficiently to restore capillarity around the seeds for better germination and early growth. A roller could be made by inserting heavy axle pins in the ends of a smooth log and attaching an animal's traces to these by ropes. An occasional farmer used a hollow roller made of planks arranged like staves of a barrel, with an axle passed through the ends.

On modern-day acreage it is not unusual to see a heavy tractor pulling gang plow, disk, and harrow in one operation. One "new" experiment, which is something of a return to the old shovel plow of the

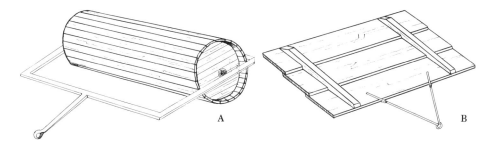

The roller (A) or the planker (B) was sometimes used to smooth harrowed ground before seeding.

deadening, is the chisel plow. It loosens the soil to considerable depth, providing aeration without actually turning furrows upside down. Perhaps the long abandoned jump shovel was not entirely bad! In today's preparation of the soil for planting, one worker does many times what an entire family could do in pioneer days.

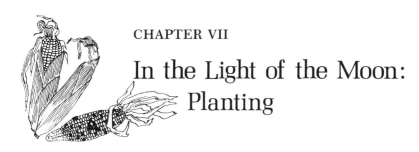

CHAPTER VII

In the Light of the Moon: Planting

When now the ox, obedient to thy call,
Repays the loan that fill'd the winter stall,
Pursue his traces o'er the furrow'd plain,
And plant in measured hills the golden grain.
JOEL BARLOW, "The Hasty Pudding," 1793

One for the blackbird, one for the crow,
One for the cutworm, and one to grow.
An early folksaying

A small cluster of farmers of Missouri's east side stood swapping plans and hopes for the spring planting. As so often happened, the subject of the moon's phase came up. Wry, slow-spoken Charlie Jeffries, one of a group of skeptics, which had grown large with the waning of the nineteenth century, bantered at the believers. "I don't plant my crops in the moon, by thunder, I plant 'em in the ground!"

Superstition played no small part in the lives of red, white, and black farmers before the late nineteenth century. Given the rather primitive state of science and the life-or-death importance of corn to both Indian and settler, it is not surprising that this crop was all but surrounded by an aura of the supernatural. Settlers scoffed at the Indians' myths, then trudged off to the fields and followed their own. Some tribes of the eastern woodland believed that crows had brought them the great gift of corn, and thus these wise birds were given special protection despite their appetites for newly planted kernels. Squaws, as the child bearers of the tribes, were believed to have a kind of monopoly on fertility. In most Indian cultures, they performed the labor of planting with the expectation that their genius for reproduction would germinate the seeds. For that matter, they did nearly all of the corn work from planting to eating time. One wonders if the fertility myth did not start as a cunning way of "farming out" the labor. Men usually had much more free time than women. Longfellow's Minne-haha in *The Song of Hiawatha*, true to the actual Chippewa practice recorded by ethnologist Henry Schoolcraft, removed her garments at night and dragged them three times around the freshly planted field to banish evil spirits. Some tribes stopped sexual activities for a few days before corn seeding, then engaged in intercourse at planting time. Indian, white, and black alike offered up prayers and incantations to their respective gods for help in pushing up healthy sprouts, bringing rain, banishing enemies, and unfolding bountiful harvests.

Most widespread of the pioneer myths about planting related to the moon. The many cultural stowaways to America from Europe included a legacy of lunar mythology with roots in ancient and medieval times. Almanacs paid homage, with tables and tips, to the magical powers of the great golden orb in the night sky, and folklore offered reenforcement. As the moon grew from crescent to first quarter to full, so it was supposed that corn, planted in this waxing moon phase, called the light of the moon, would be spurred to greater growth. Some diehard moon-cult followers even tracked about in semidarkness, dropping their corn seeds and potato cuttings by moonlight. But most farmers limited their observance to planting anytime during the calendar period of the light of the moon. Time and successful defiance by skeptics dispelled the mythology, although a few moon planters persisted well into the twentieth century. Brush clearing, tree chopping, butchering, calf weaning, and many other farm frontier tasks were geared to moon phases, but nowhere was the moon myth more tenacious than in its influence on the planting time.

That buried seeds grew into plants and reproduced their own kind was known to Indians and Europeans thousands of years before the landings at Jamestown and Plymouth. It was well understood that baby corn plants were not delivered by storks. Crows, perhaps, but not storks. Moreover, the science of seed selection was well advanced. Indian development of all six colors and six basic varieties of maize known today is clear proof of this seeding and breeding know-how, whatever their superstitions. The whites learned much from their native hosts and proceeded to improve on the techniques.

Seed selection processes were simple and obvious. The farmers used kernels from the largest ears with as little taper as possible, and with most rows of grains on each, and discarded irregularly shaped grains from the ends of ears. Evenness, plumpness, proper tightness (not too loose, not air tight), and depth of grains were important considerations. For best results, the settlers selected seed corn that met a number of other requirements, including two or more good ears per stalk, early-drooping ears with full husk coverage for protection against weather and weevils, and ears averaging no more than four feet above ground (high ears caused broken stalks and spoilage). To forestall damage from early frosts, pioneers saved the first good ears to ripen and thus gradually cut the maize growing season from 150 to 120 days in the tropics to 90 days or less in higher latitudes. A European visitor of the 1740s reported corn that matured in ten weeks in the more northerly colonies. Seed corn gatherers watched for the first drying husks in the field and marked the ears, leaving them on the stalks for full maturing before plucking them for seed.

Some settlers were very careless about seed selection, choosing the ears from the cribs without any knowledge of the stalks on which they had grown. On the other hand, there were those who set aside rich plots of ground to be used exclusively for growing seed corn. These forerunners of modern scientific growers were the most successful. The usual method of drying was to pull back the husks and leave them on the ears. Husks of a number of ears were tied together and clusters of ears were hung in dry, airy places, such as to rafters in the attic, where rodents could not climb down to them and where temperatures would stay above freezing. There they would remain suspended like brilliantly colored chandeliers until the next planting season. Some farmers used rodent-proof screened cabinets for seed corn storage, as well as nails in posts and wire spikes on which they impaled the ears to hang them out of reach of mice and rats.

Because of corn's productive gain—often as much as several thousand grains produced on a multieared stalk for each grain planted—the settlers sacrificed only small amounts of food grain by withholding seed from the harvested crop. Two bushels of seed would provide an ample year's crop for a large family. Ten pounds of grain was considered adequate for planting an acre of ground. Two or three times as much seed as required for a year's planting had to be hoarded as a precaution against drouth or other catastrophe. Seed corn was most virile for the first two or three years, although some would produce after many years of storage.

Some farmers of early America gave their seed corn a running start with special treatment immediately before planting. They soaked the grain with water, hastening germination and shortening the growing season by several days, a decisive margin in some seasons of early

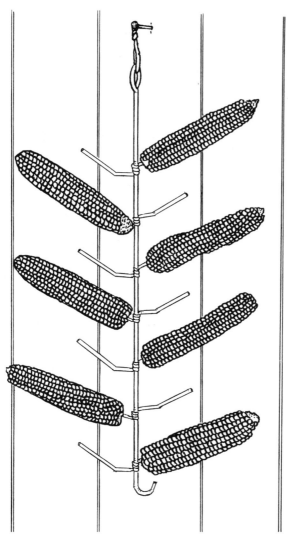

This rodent-proof seed-corn rack was made of wire. Other hangers were made by nailing spikes into posts, or ears were hung from the rafters with twine.

frost. Several devices were known to have been used to give a special fertilizing boost. Some farmers followed the coastal Indian practice of planting one or more small fish in each hole. Others would drop a handful or shovel of manure in the hole with the seed, a practice that began at least as early as the 1700s. Finally, one "Father Abraham," writing in a farmers' almanac of 1808 for the meridian belt of the middle and western states, advised boiling in water a solution of equal parts of "sheep, horse, and cow dung," cooling it, and coating the corn seeds with the odorous potion. This method, long known in China and the West, was intended to cause quick germination from the moisture and fast growth from the nutrients.

The raids of crows, maize thieves (blackbirds), squirrels, and other enemies forced other preplanting treatment of seeds. In the eighteenth century, some settlers cooked a solution of water and the root of a swamp plant called hellebore and soaked the corn grains in it overnight before planting. After eating several grains, varmints would become dizzy and tumble over, frightening the others away. This mixture did not damage the corn and the moisture hastened sprouting. A century later, some farmers were soaking their seed corn overnight in mixtures containing such ingredients as salt water, lime, ashes, gypsum, red lead, saltpeter, and even urine to speed up germination and discourage birds, animals, and worms. The most successful potions contained, as the chief agent, pine tar or coal tar. This was combined with warm water and mixed with seed corn until the kernels were a yellowish brown color. It did not interfere with germination, but even after the grains had sprouted and plants were several inches tall, most enemies found it distasteful, leaving the field alone after sampling one or two grains. By the mid–nineteenth century, a federal government agricultural advisor recommended planting corn just as it came from the cob, presumably because the pesky little seed eaters were much less numerous than in earlier years.

An English farmer on Long Island demonstrated that corn sprouted and grown two feet high in a hot house, then transplanted to fields, survived well. This practice foiled birds, squirrels, weeds and early frosts and, according to claims, took less work, but for some reason it did not catch on. Perhaps early-maturing varieties of corn made it unnecessary in growing seasons of normal length.

The season for planting, like so many other aspects of agriculture, depended upon part guesswork, part compromise, part gamble with a pinch of intuition and superstition. Ten days to two weeks after the last killing frost, one theory held. But how does one know which killing frost is the last? Fortunately, corn's hardiness would forgive some freezing temperatures shortly after planting. Moon phases, weather, and moistness of the soil entered into planting decisions. There were variations with latitude and seasons—early March in the Deep South, early April to early May in Illinois, and even as late as June for the northern fringes of the corn country. "Get your corn in by May 10" was the common advice for farmers of the corn belt from Ohio to Nebraska. Beyond that date there was a price to be paid in crop yields.

Conversely, those who challenged late frosts by early planting burned their bridges behind them unless they had surpluses of seed corn.

Aztec and Mayan calendars, probably developed either by or for corn culture, were the most accurate in the world for their time. But north of Mexico, the Indians relied on nature's annually recurring voices of spring, the nesting of birds, the size of new leaves, or the blossoming of flowers. And despite their own calendars and almanacs, pioneers often as not followed Indian signs. "Plant your corn when oak leaves are the size of squirrels' ears" was the most universally adopted rule. There were other rules—whenever the broad, white petals of the dogwood were fully unfurled, "when the apple is bursting its blossom buds," and among the Hopi of Arizona, "when the rising sun was aligned with fixed landmarks on the eastern horizon."

Another ancient Indian custom often followed by settlers was to stagger the planting times of different fields to spread out cultivating and harvesting work and lengthen the fresh corn season. Practically all sources contend that no more than one corn crop could be grown on a given piece of land during a growing season, although, in 1822, a frontier experimental horticulturalist near Franklin, Missouri, reported raising two good corn crops on the same Missouri River bottom clearing in one year.

Just as views differed about the time of the year and the phase of the moon to plant corn, so there were variations as to widths of rows. In porous, sandy desert soils such as those of the Hopi, Navajo, and Zuni Indians of Arizona and New Mexico where corn roots reached to great distances, it was not unusual to plant hills six to twelve feet apart. The plowhands who crosshatched their fields in the Northeast of the 1740s laid out rows in checkerboard squares from four to nine feet part. Indian corn did not require the widths that early settlers often gave it between rows, but wide spacing was traditional, and tradition was backed with a fullness of frontier beliefs and practices. High on the list of reasons was, as William Cobbett expressed it, "room for good, true, and tolerably deep plowing" between rows. A width of five or six feet allowed for driving animals between rows, both for plowing and later harvesting with carts. Then there were farmers who planted other crops between maize rows, and narrow spaces would have let in too little light. Thomas Jefferson had his slaves plant potatoes between the lines of corn. Many southern planters spaced their corn rows five and one-half feet apart because that had been the practice with cotton. About 1840, Martin W. Philips and A. K. Montgomery of Mississippi doubled their yields by planting rows only half that distance apart, and drilled rather than squared. The term *drill* apparently came from the military drill step of about two feet. Drilled corn was planted anywhere from two feet down to six or eight inches apart in the row. It was at least 25 percent more productive than hilling in squares.

Those quiltwork squares, the warp and woof of rural America laid out with great labor and care to facilitate cultivating both lengthwise and across the field, were beautiful to see. The farm family with the neatest, straightest rows running in both directions took great pride in its work and was the envy of the neighborhood. The resulting diagonal lines or avenues too were straight and pleasing to the eye. It had been done this way by the Indians for untold centuries. And for nearly 350 years after the founding of Jamestown and Plymouth most American farmers would continue the practice. Yet it was not merely useless; it was worse than useless. It required far more time to lay out and plant, and as a sizeable number of eighteenth- and nineteenth-century farmers well knew, it produced less corn than the drill pattern. Finally, the two-way cultivation, a basic reason for the pattern, caused much greater soil erosion, negating the benefits of contour farming. Such is the power of tradition that checkerboard planting persisted on many farms to the middle of the twentieth century.

Corn planters laid out their checkered row patterns in various ways. In the early 1500s, Oviedo observed that the Indians would walk perhaps six abreast, simultaneously step forward, plant at each step then repeat the process, keeping in line until they had crossed the field. One popular early settler's method was the crosshatch plowing with grains planted at the furrow intersections. An ox-drawn log, pulled endways back-and-forth, then criss-cross, was another technique. Some farmers drove their teams with sleds up and down lengthways, then back-and-forth crossways, leaving two shallow trenches with each trip across the field. Hamlin Garland tells of using four-track sleds on western prairie lands. Grain was then planted at intersections of all lines left by the logs or runners. Another method was to drive clearly visible peeled white stakes at the ends of the field and walk directly toward each, or to plow a shallow trench with a hand-pushed wheel plow. At the end of each crossing, a stake would be moved over the width of two rows and the field would be recrossed to the stake at the other end. Until the late-nineteenth-century invention of machines that simultaneously marked and seeded in checkerboard rows, the neatest planted rows were those laid out by the crisscross sled run-

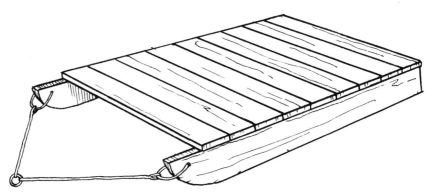

Sleds like this were used to mark crosshatching on the plowed and harrowed fields for planting, and they often doubled as stone boats to haul rocks from the fields.

ner markings. Sometimes general utility farm sleds and stone boats doubled for this purpose.

Without a doubt the easiest and fastest part of the entire corn growing operation, despite its back-bending nature, was planting the seed kernels. The land was cleared, except perhaps for the leafless trees of a deadening, and crosshatches were marked. Wise farmers followed the Indian practice of testing their seed for germination several weeks before planting time to be sure it had not been damaged by freezing. Women, children as young as six years, and men marched like small platoons down the rows, each one counting out grains from the sack hung over the shoulder or from a can or tin cup and dropping them at the desired spots. Anson Graves related that, as a boy on an Ohio farm in the mid–nineteenth century, he planted corn two rows at a time from a tin bucket fastened by a line to his neck. He was small and nimble and could plant as fast as two men. As in many areas of agriculture, corn planting was a family affair. The Indian method, adopted for a time by settlers, was to poke holes with sharpened sticks, deposit the grains, cover, and step on the plantings to pack moist soil around the seeds. By far the most common pioneer method before 1850 was to drop the seed at cross-marks and cover with hoes. Where cross trenches had been plowed for seeding, it soon became necessary to plant each cluster or hill of seeds with the grains an inch or so apart to avoid root crowding.

How many grains to the hole or hill? An old rhyme, its origin lost in the hazy antiquity of folklore, offers a clue: "One for the blackbird, one for the crow, one for the cutworm, and one to grow." Four to five was the usual number, and tillers hoped that two to four stalks would grow. They could always thin excess stalks from each hill, leaving the two or three healthiest. They would also make a quick once-over through the field after the blades were two inches high to replant those hills where no sprouts came up.

The smaller bread grains, wheat, barley, rye, and oats, could be sown by scattering broadcast and covering lightly by a harrow or drag board, but corn was a grass of a different breed and need. Maize thieves and crows would have picked the field clean of its bright kernels in a matter of hours. Shallow planting was best for quick sprouting and growth but worst for survival in the presence of myriads of birds that would pull the sprouting blades and eat the grain. Roger Williams, Rhode Island cleric and colonizer, reported that the Narraganset Indians planted their maize grains up to six inches deep to thwart those sacred black thieves, the crows. Settlers had no such feelings of reverence for the winged invaders, and they shot or shooed them away, tar-treated their seeds, used both live and dummy scarecrows, and planted their corn three-fingers-width deep. One and a half to two and a half inches became the most common depth. Regardless of planting depth, the roots branched out about one inch below ground. Planting deeper than two or three inches created a germinating problem since corn does not germinate well below sixty degrees Fahrenheit. "This generation is very sure to plant corn and beans each year precisely as the Indians did centuries ago and taught the settlers to do, as if there were a fate in it," wrote Thoreau in the late 1840s. The Iroquois and other tribes planted beans in the same hills with corn, but several weeks later. Probably neither ethnic group had a widespread understanding of the reasons that these two crops, when grown in combination, were so valuable. Corn and beans provided one of the best examples of complementary crops. Bean vines climbed the corn stalks and yielded better harvests because their leaves and pods were off the ground. The high protein content of beans supplemented corn diets in a very important way, making possible very heavy corn consumption. And soils were more productive for a longer period owing to the nitrogen which beans added to the soil.

Between the corn hills, the Indians planted their native American crops, squashes, pompions or pumpkins, and sunflowers. Squash and pumpkin planting among the corn was copied by the white and black settlers. The resulting fruits were used as both human and stock feed. Golden pumpkins among stately cornshocks have won a secure place in the poetry, literature, art, and folklore of America.

The old hole, hoe, hatch-mark, and hand-drop systems prevailed with little challenge for over 225 years after the first arrival of English settlers upon the scene, but eventually time and technology overtook these primitive methods (although some Indians of the Southwest still use the planting stick). The second quarter of the nineteenth century saw the invention of more agricultural implements on a much higher technical level than had all the previous history of humankind, and a very large number of these innovations were American. There were improved plows, harrows, cultivators, seeders, reapers for small grains, hemp breakers, and grain threshers and shellers. What caused this sudden burst of inventiveness? Among the many and complex reasons were creative vigor from feedback of ideas nurtured by the industrial revolution; rapid growth and large food demand of urban industrial labor populations, which spurred farmers to expand from single-family subsistence tilling to commercial production; the extension of American farming to western prairies and plains, which served as a challenge for larger production units; and the stimulus of prizes and recognition offered at American agricultural fairs for labor-saving farm inventions. For example, the New York State Fair at Auburn in 1846 gave prizes for corn cutters, shellers, cob and grain crushers, and planters. Such devices did for the commercializing of various crops what ginning mills had been doing for short staple cotton since the 1790s.

Among the multitude of these nineteenth-century American creations were many devoted to the production of Indian corn. We have noted the improved plows and harrows. Of much more intricate design were the diverse types of corn planters. Shortly before the middle of the nineteenth century a simple mechanical planter came into use. Sometimes called jabber or jobber, it consisted of two narrow boards with handles at the top and a hinge near the bottom. When pulled apart at the top, the boards tripped a mechanism that metered several grains of corn from an attached seed tank. These kernels fell and lodged between two metal blades, which were closed together at the bottom ends of the boards. When the farmer jabbed the blades into the soil and pushed the two boards together by the handles, the bottom blades separated below the hinge, dropping the grains of corn several inches deep into the ground. The worker then removed the mechanical planter, stepped on the hole in order to pack earth around the

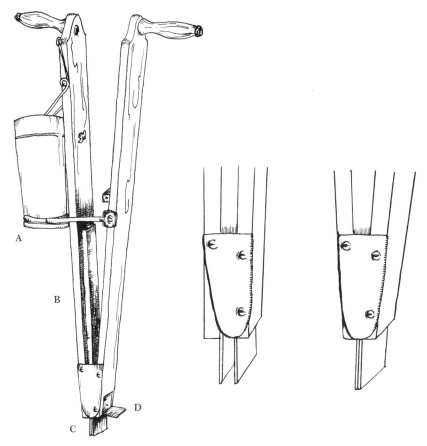

The jabber was the most popular hand-operated corn planter ever devised. Pulling the handles apart tripped the seed-metering device at the bottom of the tank (A). *The seeds would fall through the tube* (B) *to be held at the bottom by jaws* (C), *which were jabbed into the ground to the depth of the shoe* (D). *When the handles were pushed together, the jaws would separate, releasing seed several inches deep in the soil. Inset shows jaws closed for jabbing* (left) *and open for depositing kernels* (right). *The planter was also used for beans, squash, and pumpkins.*

seeds, and moved on to the next intersection of planting markers. This ingenious device, soon improved, was effective, cheap, and much faster than the hoe. In time it came into almost universal use, putting an end to the stoop labor in planting.

Similar seeders which could be used one in each hand were devised. Then came the Randall and Jones model which consisted of two hand planters fixed side by side to a board that was as long as the distance between two rows. Such contraptions enabled one worker to plant two rows at a time without the aid of a horse or mule team. However, they were not highly successful since it was difficult for the worker to step on the plantings in both rows. Too, women and children, who did a large share of the planting, found them hard to manage.

The first known wheeled, single-row mechanical drill was recorded by Deane of Massachusetts in 1790. This device plowed a row, planted, and covered in one operation, "and all this at the same time with great expedition." Not with great enough expedition, however. Single-row planters, one by A. H. and L. Robbins of western New York, and another by Charles R. Belt of Georgetown, Maine, were invented in the 1820s. "Awkward," and "unreliable" were terms used to describe the wheeled planting machines of this time as well as others developed in the following decade.

By the 1850s, inventors had greatly improved mechanical planters. Single-wheeled models manufactured by a number of small producers, including Billings, Pratt, Cole, Emery, and Harrington were catching on. Some of these were pushed by hand, like wheelbarrows, while others were pulled by animals. Billings' machine was equipped to drop concentrated, dried manure in the furrow with the seed. Topping all these by a wide margin, Brown's two-wheeled, two-row planter could seed twenty-five to forty times as fast as one worker with a hoe.

Nevertheless, by the 1860s, a majority of American farmers were

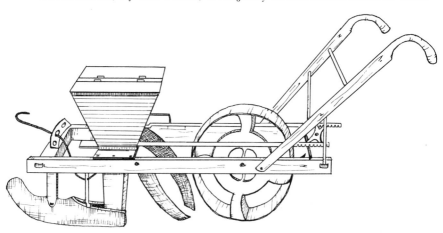

The Billings single-row, horse-drawn corn planter of the mid–nineteenth century deposited dried fertilizer with seed corn.

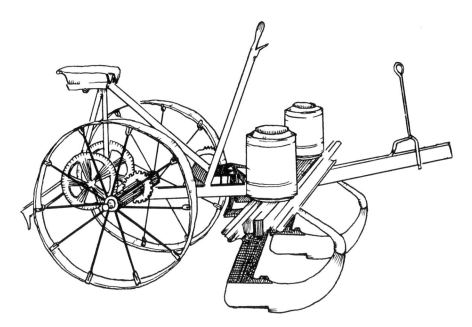

In this two-row, horse-drawn corn planter of post–Civil War design, seeds in the cylindrical tanks were metered to shoes below, deposited several inches underground, and tamped by the wheels.

still relying on jabber hand planters and occasionally hoes. The reasons were several: the availability of child and woman labor, the expense and breakdowns of machines, and a tradition-locked conservatism which steadfastly insisted on checkerboard planting. But the mass mechanization of America following the Civil War reached out to the prairies. Rail lines increased access to markets, commercial production grew by giant strides, and hand planting could no longer meet the challenge.

Fundamental to the acceptance of the two-row, horse-drawn machines was the development of a reliable means of planting in checkerboard rows or squares. First a rider, often a child, seated with the driver on each machine, pulled a lever which simultaneously dropped seed corn from two tanks through holes in the metal shoes running several inches deep in the soft, harrowed earth. The lever was moved as the machine crossed over each sled mark on the field. Concave wheel rims rolled over the seeded grooves and crimped the soil tightly enough to enable the kernels to germinate readily. From this advance it was but a step, accomplished in 1875, to stretch a cord along the field and drive the team of animals along it. Knots every three or four feet (eventually standardized at three feet eight inches) on this line tripped the planting mechanisms, seeding both rows at the same time. After each "run" across the field, the driver stopped and shifted the cord one row-width toward the unplanted area of the field. A marker arm on the machine dragged a straight line parallel to the planted rows. Good draft animals quickly learned to straddle this shallow trench or groove with each trip the length of the field. Several years after the invention of this cord-tripped planter, a reel of wire with small wire knots every three feet eight inches replaced the cord. In the 1890s, the "single kernel accumulative" counting mechanism permitted planting the precise number of grains desired in each hill. Actually, one-way cultivation caused less erosion of soils since it was consistent with the principles of contour farming.

Those picturesque, symmetrical, checkered corn fields of centuries past are another part of the vanished American scene, save for their existence in the folds of picture albums. Today's planters are set to drill the seed corn at intervals as short as eight inches, in rows two and one-half feet apart, or less. Much more efficiently than Billings' one-horse planter of the 1840s and 1850s, they deposit balanced kinds and carefully measured amounts of fertilizer on the fields. At the same time, the modern mechanical seeders administer weed killers and eliminate much of the toil of cultivation that characterized corn growing throughout the pioneer period and up to the mid–twentieth century. Had such herbicides existed by the 1600s, the succeeding chapter would have been unwritten.

CHAPTER VIII

First the Blade:
Cultivating

He that neglects the weed
Will surely come to need.
Old Farmer's Almanack, May 1831

The poor corn plant, if left to itself, will soon be like
Gulliver, when bound down by the Lilliputians.
William Cobbett, 1828

Thrice in the season, through each verdant row,
Wield the strong plow share and the faithful hoe.
Joel Barlow, "The Hasty Pudding," 1793

Corn and hoes for three hundred years were as closely linked as corn and hogs. But no longer. Today's corn farmer has about as much use for a hoe as for an Indian planting stick. Weed killing sprays have revolutionized corn growing in the last thirty years. Most modern farmers use herbicides at planting time, which give the corn a big head start. Long afterward they may cultivate once to kill late weeds, but the trend is toward "low-till" or "no-till" growing. In the latter case, herbicides do all the weeding and there is no cultivation at all. Like the ox or horse in the furrow, however, the place of the hoe in American history is secure.

As that brown field of plowed, disked, and harrowed earth sprouted sprigs of corn, it looked for all the world like a giant piece of graph paper, its squared green lines stretching from border to border. Bypassing the varmint menace for the moment, we find that the infant crop of corn was faced with two problems—water and weeds. Following rain, the fields, it was believed, had to be mulched or broken to prevent evaporation and hold the moisture in the soil. Further, the settlers thought that they had to loosen the dirt around corn plants by cultivation three or four times during the growth period in order to aerate the soil and enable it to absorb more moisture, and to allow the

Indian hoe.

roots to penetrate better. Modern agriculturalists have demonstrated that it is preferable to leave the roots undisturbed and the soil unbroken except as a last resort to kill the weeds not checked by herbicides. But even if they had known this, the pioneer farmers could not have done much differently from what they did. Lacking herbicides, they had to hoe and cultivate or lose their crops to weeds.

When the corn was two to six inches tall, tillers could "dust mulch" a rain-packed soil by harrowing without doing significant damage to the plants. A youngster could be instructed to follow the spike-tooth harrow and "set up" the few sprouts disturbed by harrow teeth. Harrowing was best done by sunlight, since tender corn plants were less likely to break when warmed by the sun, and weeds died quickly in hot sunlight after their roots were loosened and exposed by harrow teeth. As the tender shoots grew, harrows gave way to hoes and animal-drawn plows and cultivators. For countless years hoes were all but indispensable to corn culture, as necessary as corn was to human existence. Indians had used them for ages, making them from antlers, seashells, wood, flint, and the scapulas or shoulder blades of large animals, with stick handles lashed to them. Colonists sometimes copied these primitive tools, but usually they brought their own iron and steel hoes with them to the New World. These were all-purpose implements. They could turn sod like spades, break and smooth the soil,

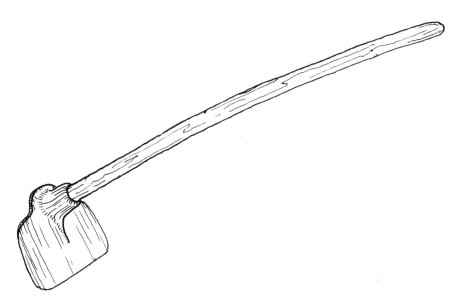

*Settlers used this metal hoe for chopping weeds, backsodding, and with the
handle removed, for baking hoecake.*

chop and cover for seeding, hill up or trench around plants, and dig
irrigation canals. When corn was too small to risk cultivation with
draft animals, hoes were ideal, as they were when the stalks were too
tall to be tilled by cultivators without breaking or damaging. Several
hundred years after horse-drawn cultivators came into wide use, farm
families were still extensively employing hoes in corn fields.

Sometimes because of illness in the family, sickness among livestock or muddy field conditions after heavy rains, the farmers fell behind in weeding and their crops were said to be "in the weeds" or "in
the grass." Plows and cultivators were not always adequate to handle
the job, and hoes, slow, back-breaking, but efficient, were the old reli-
ables. Hard labor it was indeed. John Muir of Wisconsin branded corn
hoeing as "deadly heavy" work. Despite the advice of various sages, to
the effect that "it doesn't pay to hand-hoe corn" except in the South
where "droves of slaves" cover the field, millions of acres were hand-
hoed, and it did pay. In the corn patches of New Mexico, hoes gave
way to cultivators much more slowly than in the East and Midwest.

Pioneer families were often large and it was a common occurrence
to see a line of workers, man, woman, and boys and girls of various
sizes, hoeing in parallel formation down the long corn rows. Their implements had been taken fresh from the "rain barrel" where overnight

soaking had swelled the hickory handles to prevent loosening of metal ferrules. Oldsters, guns strapped to their backs for protection or shooting game, would take two rows so that children could keep apace. Jefferson stated that a good hand with a hoe could weed five hundred hills of corn in a day. But Sam Steele of the Shaker colony hoed corn so slowly and stood so long in one place that, according to a jesting observer, the shade from his brimmed hat killed the corn plants. Tedious and boring, yes, as the hacking proceeded with glacial slowness down seemingly endless rows, under the hot, midsummer sun, made more stifling by high corn which blocked what little breeze was stirring.

Chop, chop; cut that horseweed, fennel, and pigweed; mulch that soil; no telling when it will rain again; guard those corn and bean and pumpkin roots; hill up the stalks at the base; hack out the ragweed, the crabgrass, and smartweed. And don't touch your eyes after handling smartweed. Stop and pull the bandana from the hip pocket and mop the sweat of honest labor from the brow; slap that fly and the tenacious little stinging sweat bee; pull the well-worn whetstone from hip pocket and give the hoe blade a few rasping strokes to sharpen it for the remainder of the row. Toss a clod of dirt occasionally at an unsuspecting co-worker to lighten the spirit of the drudgery, or call time out for a joke or a story to mingle amusement with a moment of rest. And race, perhaps, to see who will get to the end of the row first. There is rest ahead, where leafy limbs of trees bordering the field dip down to meet the growing blades of corn. Rest in the cool, deep shade, and take a refreshing draught of water from the jug. There is a mood, too, of satisfaction, of contentment that, but for weariness and sweat, would border on sheer delight at looking back on clean rows. There is pride over the feeling of being a needed member of a team and achievement born of the realization that those rows of corn were "laid by" until the harvest of bright autumn days.

"Laid by" was a corn term, meaning that settlers had finished weeding and mulching, and that the corn, still short of silking and tasseling, was tall enough to outdo its leafy competitors and to mature with no further work. The expression applied to tending with horse, mule, or ox and cultivator as well as hoes. Draft animals were the greatest labor savers for pioneer settlers as compared with Indian methods of farming. By the 1740s, many American farmers were cultivating corn with the use of animals.

Settlers disagreed over which animals were superior. It was a "hare and tortoise" contest with advantages on both sides. Horses and mules were unquestionably faster, but sometimes their sudden lurches broke traces or whippletrees. Oxen were less responsive to verbal commands, and their feet were less durable on hard roads. The ox, however, had its strengths. It was not as temperamental as the horse, was less susceptible to illness, cheaper to feed, and less particular about eating, requiring no bins, stalls, or chaff cutting. Furthermore, it required no shoes and a minimum of foot care. The ox cost only about half as much as a horse and when not working, could be fattened for meat and sold for as much as or more than the purchase price. And it was a strong and steady worker. Swearing *by* one's work animals was as customary as swearing *at* them. Cultivation with animals could begin when the young plants were six to eight inches tall.

Corn farmers had perhaps more divergent opinions about cultivating than about any other aspect of corn culture. "Plow shallow"; "plow deep"; "plow often"; "don't disturb the roots"; "hill-up the stalks"; "don't hill-up"; "plow less and hoe more"; "hoeing is a waste of time for any farmer with a horse or an ox." Such was the confusion of beliefs with which every farmer had to contend. Corn hoeing and plowing occupies much space in the early almanacs and diaries. And such was the tenacity of tradition that little was changed by intelligent experimentation and accurate reporting.

Tradition probably designed the first American corn cultivating plow. It is not surprising that this device looked very much like the

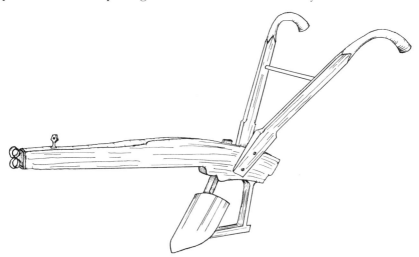

The single-shovel cultivator required two trips the length of the field for each row of corn.

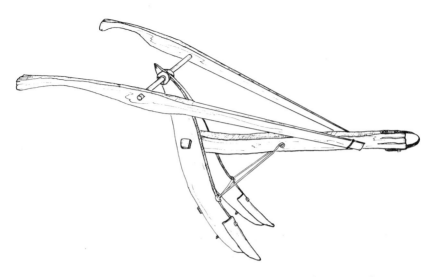

This double-shovel cultivator of early to mid–nineteenth century design was the most widely used horse-drawn cultivator. Only one round trip per row was required.

jump shovel plow used for the first breaking of land in a new clearing or deadening. Sometimes plow hands relied on the same plow for cultivating. Usually, however, it was a smaller version with much the same lines. This "single shovel," or "horse hoe" required two round trips (four times the length of the field) between each row.

Then came the most popular one-horse cultivator ever invented, the double shovel, which required only one round trip for each row. It was never rendered obsolete on the small farm until cultivation itself became practically outdated. As late as the 1940s and 1950s the old reliable double shovel was still in use. Nearly every farmer kept one in the tool shed or lot as a backup in case one of the animals became sick.

Through the colonial period and up to the 1830s, deep plowing between the rows was the most popular advice offered to farmers. As late as 1830, the *American Farmer* carried a recommendation to plow nine inches deep on the last cultivation. Opinions had been changing, however. Between 1775 and 1812, implement makers designed one-horse cultivators with five to a dozen "shoes" or shovels. They ran shallower than the single and double shovels. During the 1840s, double shovels began to give way more rapidly to two-horse cultivators which had more but smaller shovels. The theory, and an accurate one, was that deep plowing tore the roots of corn plants and damaged growth. These cultivators were a distinct improvement. Not only did they run shallow, but they cultivated two rows for every round trip, twice as fast as the double shovel.

Designers soon equipped new models with wheels. (The first patent for such a machine was issued in 1846.) The two "banks" of cultivators were flexible and operators could pull them closer or farther from the plants by handles; or, in the case of riding models, with stirrups, controlled by their feet. A metal fender on the inside of each bank of shovels kept the loosened dirt from covering the small plants. Disk cultivators, patterned after disk harrows, were developed in the late nineteenth century. These were usually of the riding types. As in the case of plows and harrows, many farmers looked with disgust upon anyone who would ride a cultivator.

Still another cultivator design, the sweep plow, emerged at the end of the pioneer period. It had a series of V-shaped weeding blades which made shallow cuts. These cutters disturbed corn roots less than the shovel models and, because they pulverized the surface much less, they reduced blow-away erosion in dusty, high-wind areas of the plains.

The reason settlers continued the Indian pattern of planting corn in checkerboard design, as noted earlier, was to permit them to cultivate

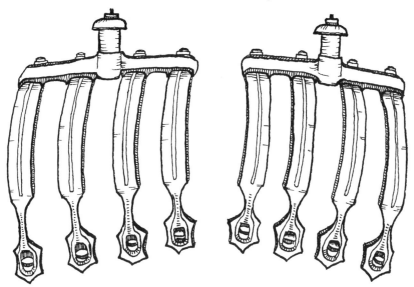

The two-wheeled cultivator with its two banks of multiple shovels (shown above) *was twice as fast as the old double-shovel.*

lengthways one time, and crossways, or at right angles, the next time, thereby more effectively uprooting or covering weeds. The crisscross cultivation practice had been brought to America from England where experimental farmer Jethro Tull introduced it in the early 1700s.

Another Indian method was to hill up the soil around each plant. Hence the expression, a *hill* of corn or beans. Settlers for generations faithfully followed this example, believing that it braced and strengthened the stalks. By the 1830s, however, some agricultural experts advised that hilling prevented aeration and caused erosion that exposed the roots of plants, retarding growth. In 1871, Horace Greeley wrote that, as a boy he "hoed diligently for weeks at a time, drawing the earth up about the stalks," or "hilling up," only to learn later in life that this had shortened root growth and made the plants more vulnerable to drouth. Thereafter, some farmers tended to keep their cultivators from throwing too much dirt against the stalks, but hilling or ridging with horse-drawn cultivators continued on some farms until the mid–twentieth century.

Horses and mules had excellent "horse sense" and would follow down the desired rows unerringly. But there were occasional problems. The cultivator operator, as did the plowhand, had to sidestep animal excrement on occasion. More annoying was the fact that the oxen, horses, and mules would incessantly lower their heads, drop step, and catch bites of the tender, succulent corn plants while plodding down the rows. This caused both an irregular gait and damage to the young plants and could not go unchecked. Teamsters stopped it by placing a leather or wire muzzle in the shape of a basket over the nose and mouth of each animal so that it could breathe but not eat. The beast presented a comical picture of frustration as it dutifully walked between the rows of tempting green morsels, blinders behind its eyes, a muzzle over its nose, and a bit in its mouth. Sometimes drivers wrapped bits with wire, the better to guide tough-mouthed, unresponsive animals.

There were many other tasks that tied the settler to the corn crop between sprouting and maturity of the stalks. Frequent rains meant rank weed growth and more hoeings or plowings. Hail storms were the nemesis of the crop, shredding the leaves and scarring the ears. Drouth brought anxiety, prayers, and "firing" or drying of the lower leaves if the dryness was long continued. A multitude of pests tested the farm family's patience and ingenuity, as will be seen later.

All hands turned out to thin the corn when it was six inches to a foot

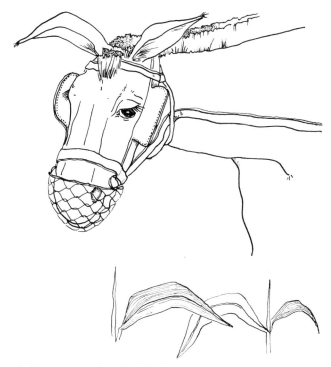

Mules were muzzled to prevent their freeloading on young corn plants during cultivation.

in height. Where more than three grains came up in the same hill, thinners pulled the smallest of the excess to prevent overcrowding. The rule of many southern farmers was to keep only the two healthiest stalks in each hill. A few weeks later as the plants neared their full height, suckers would grow outward from the bases of some stalks. Indians had pulled these off from time immemorial, and European and African settlers and their descendants followed their lead down to modern times, believing that these offshoots drained or diverted nutrients from the ears. Children did much of the thinning and pulling of suckers. In recent years some agronomists have claimed that suckers do no harm and should be left on the plants. Where corn is cut for silage, suckers add to the tonnage of the harvest.

It was a hauntingly beautiful and romantic time, this stage of the maturing corn crop. Moonlight glowed on a thousand swordlike blades, reminiscent of an ancient army silently poised for battle. Rhythmic whines of cicadas or "great harvest flies" in late afternoon yielded at dusk to creak and rasp of crickets and katydids; and whole fields after dark were electrified with the tiny flashes of fireflies, better known as lightning bugs. A night breeze stirred and the lances waved in defiance. Farmers agonized over parching weather and watched the distant heat lightning as storms failed to develop or passed them by. Under ideal conditions corn grew as much as four and one-half inches in a day. And on warm nights when breezes died and whippoorwills, screech owls, and nighthawks ceased their eerie calls, one could stand amidst the sea of stalks and hear the corn grow. It was but a faint rustle, and the sound was not directly from growing, but from a leaf here and there suddenly uncurling after a fraction of an inch of growth. Like the nameless sounds and sights and smells of earth all about, it conveyed a feeling of oneness between nature and human. The toil had been hard but the land was bringing forth that promised "increase" as a reward. With luck and pluck, the harvest would fill crib and bin and barn before another winter gripped the scene in its icy embrace.

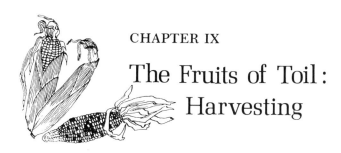

CHAPTER IX

The Fruits of Toil:
Harvesting

Harvest Indian corn without delay or birds and squirrels will.

Old Farmer's Almanack, October, 1800

And the corn, oh, there was no end to that. There were several barns, some big and some little, but when the corn was gathered and the "corn-shucking" was over and the crop was housed, the barns were full to overflowing. They would remind one of Pharaoh's barns in Egypt at the end of the seven years of plenty.

IRVING LOWERY, *Life on the Old Plantation in Ante Bellum Days*

Very little about agriculture, particularly in America's pioneer stage, was definable or controllable in mathematical terms. Down on the farm, one could not turn off the power, lock the doors of the plant, send the workers home, and reopen after a three-day weekend. The growth and ripening forces and, at times, the destructive power of the seasons moved relentlessly onward. Sometimes the earth's "increase" was a decrease. Nevertheless, harvest, of all pioneering experiences, was one where the whole was greater than the sum of its parts.

No dictionary could ever convey the full meaning of the word *harvest*, for no words have been coined to form a fitting caption to the harvest scene. It was the reaping or gathering of a crop, yes; but also much more. It was a blend of gathering and hauling, of feelings of achievement and self-reliance, of the sight and smell and feel of fruits on stalks or trees or vines. The harvest was an emotional experience as memorable as it was indescribable. Trees were changing their green mantles slowly to red and purple and gold, outlined against the blue autumn skies. The harvest moon would tone down these colors to varying shades of gray with black shadows underneath, as workers often labored on into night to foil the coming frost. The lonely calls of wild geese and brant would resound as the wedged banners of their south-bound flights were seen, in the words of William Cullen Bryant, "darkly painted on the crimson sky." Apple cider dripped from the press and yellow jackets winged down for a sip. Potatoes and turnips were buried in deep earthen trenches for winter use. Best of all, pent-up energy was at last released. The toil and worry had at last been rewarded. It had been a good year for the crops, and with careful management the harvest would see the family through another of those ever circling dormant seasons when crops marked time till spring.

If there was one typical harvest month above all others, without question it was October, the month of "bright blue weather." Perhaps Aldo Leopold, writing shortly before his death, caught the flavor of the season best. "I sometimes think that the other months were constituted mainly as a fitting interlude between Octobers." Harvest was a season, but it was more than a time of year.

The very word *harvest* was so closely linked with one pioneer crop that a New England dictionary of farm terms (1790) defined it thusly: "HARVEST, the season when corn is cut down and secured." It was typical of such a versatile crop that there were many harvests, and variations of harvests. Today's combine, or one-operation picker and sheller has simplified, if not dehumanized, the process, but there is a rich and diverse heritage in the old corn harvests that became a part of the American fiber and character. Most crops were harvested when their fruits were ripe, and that was that. There were different times of ripening of beans, squash, pumpkins, cucumbers, apples, and peaches, but they were as simple to harvest as corn was complex.

Corn harvests by the Indians too were quite simple. Some tiny, immature ears, no more than two or three inches long, were plucked and shucked to be eaten, grain, cob, and all, as vegetables. Indians did much of their harvesting when the ears were full-grown but with soft, milky grains. Later they gathered those ears which had been left to dry on the stalks so that the hard grains could be pounded into meal. Last, the native women, upon whom most of the farm work depended, pulled, piled, and burned the stalks, leaves, and tassels in preparation for the next season's planting. Many tribes, including those of the eastern woodlands, stretched out the harvesting process by intentionally staggering the planting schedule for different fields. This had the advantages of spreading both the work and the benefits, since they could enjoy fresh corn almost continuously over several months.

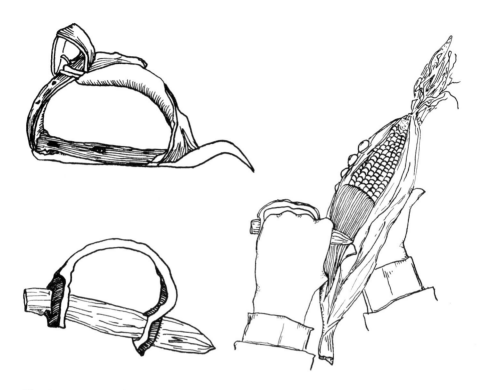

Shucking pegs and how they were used.

Somewhere along the pathway of the centuries, Indians of the eastern wooded areas devised a simple but clever little harvest hand tool which was so basic and perfect in design that the settlers adopted it and used it for 350 years with almost no improvement. They gave it the name of husking or shucking peg. A five-inch piece of hardwood, usually hickory or oak, was sharpened at one end and gripped in the palm of the hand with the sharp end next to the thumb. The peg, sometimes called a pin, was held securely to the hand by a leather or rawhide strap that passed over one finger, or perhaps several fingers. The user grasped the husk between the thumb and the sharp end of the peg and tore it loose from the ear. Pioneers eventually substituted a piece of deer antler or metal for the wood peg and experimented with a few variations in design, such as the palm hook and thumb spurs, but wood was cheaper and more accessible. Charring the point made it wear better and kept it from softening with moisture. The basic Indian design is still in use though rendered nearly obsolete by mechanized pickers. The shucking peg was one of the first of a number of peculiarly American tools related to corn culture.

Settlers too spread their plantings over the spring and summer calendar to even the work loads, but here their corn culture parts company with that of the Indians. One of the many differences was the farm animals of the pioneers—animals that had large winter appetites. The settlers thus had to grow far more corn per capita than did Indians, but they had the advantage of horse power and metal tools to do the job. In other words, horses, mules, and oxen made much larger corn crops both necessary and possible.

In a sense the harvest began when pioneer farm families thinned their corn and pulled off the suckers, since they fed these like hay to the livestock. As did the Indians, settlers harvested some baby ears, either pickling them or boiling and eating them as tender vegetables. The soft, roasting ear (or "roas'n' ear") stage was also relished by the immigrant farmers and their descendants. In season this was a staple, as will be seen later. From this point the harvest entered upon a stage of much greater variety. Unlike the Indian processes, pioneer methods wasted little, since fodder was necessary for the farm animals.

Providing green feed to livestock during the winter months was a problem. Few grasses survived the cold, but some corn farmers found an excellent alternative. It was called ensilage, shortened to silage, and was stored in silos. When the stalks were full height and the lower leaves were beginning to turn brown (ten or twelve days before ripe cutting time), the family of farm workers large and small went into the fields earmarked for silage and cut the stalks near the ground with corn knives. The earliest pioneers cut their corn crops with any available large knives, even swords and sickles. In time, they developed two specialized designs of corn knives. The first of these was half knife–half sickle, with a curved blade twelve to fifteen inches long and a handle of about equal length. Later they devised the modern corn knife with six-inch handle and blade about eighteen inches in length.

The green corn was tied in bundles and later hauled by wagon to the site where it would be stored. There it was chopped into small pieces with a heavy knife or a hatchet on a chopping block which was a stump or a log section. Around the mid–nineteenth century a special stalk cutter with a feeding chute and a hinged knife like a modern paper cutter came into use for chopping both silage and dry fodder. Although it could be operated by a single laborer, usually one worker

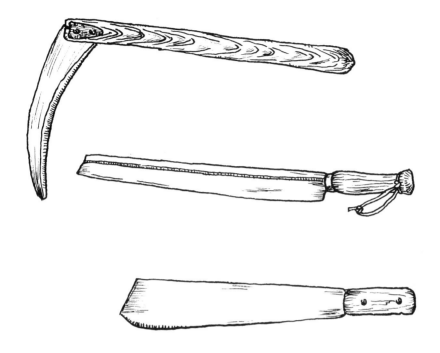

Knives used for cutting corn stalks: top, *eighteenth- and early nineteenth-century design, a modification of the grain sickle;* center, *rib-bladed knife of the mid–nineteenth century;* bottom, *post–Civil War design.*

would feed the stalks through the chute as another rapidly lifted and chopped with the guillotinelike blade. Fingers beware!

Sweating workers dumped the chopped stalks, leaves, and ears into the pit or trench, which in early times was a hole dug in an elevated, well-drained piece of ground and often lined with stones, cement, or tightly fitted wood slats. About 1875, Professor Manly Miles of Michigan and Francis Morris of Maryland built the first above-ground silos in America, round, tall towers that resembled huge barrels in both shape and wood stave construction. Often silos were erected on top of pits so that there was a continuous cylindrical storage space below and above ground. It was necessary that the chopped silage be moist, about 70 percent moisture content being ideal. Cutting time was guesswork, to be sure, but experienced farmers were good guessers as to the proper stage of maturity, and if necessary, they allowed the stalks to dry for a day or two after cutting.

When the silage was packed so that no oxygen could make its way into the mass, the growth of bacteria and formation of acids would proceed to an optimum point of partial fermentation over a period of several weeks. The silage would then remain in a succulent, edible stage for livestock for an indefinite period of time and was far superior to dried hay and fodder as winter feed. But compressing was the secret, and probably the most exhausting task in the entire corn culture was tramping silage. Broad-brimmed hats and cloths around necks were rarely enough to keep small, scratchy particles from getting inside clothing. The tramp, tramp, tramping, to press out all the air spaces was worse than running in loose sand and required frequent changes of work crews. Not to mention the fact that silage cutting time was usually in the heat of the year and there was little air circulation in the silo.

America has been called the land of silos. With the building of these

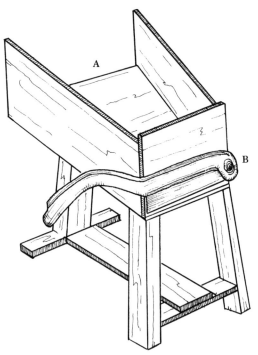

Grant's fodder and silage chopper, which worked like a paper cutter, replaced the hatchet-and-wooden-trough technique described by Jefferson. Stalks were fed through chute (A) to hand-operated cutter blade (B).

round towers in the late nineteenth century, a problem developed. The wood structures, having no internal bracing, soon took on an appearance of the Leaning Tower of Pisa, and in fact many of them tumbled over, silage and all. Builders partly remedied this by erecting the silos close alongside barns for firm anchorage. Later they constructed silos of masonry or concrete and filled them with silage chewed up by power-driven shredders and blasted into the structure through fan-driven air pipes. The silos of today are shiny, air-tight fiberglass shells that are vacuum-sealed after they are filled. Affording up to twenty tons of rich feed per acre, silage was not an insignificant product for carrying stock through the winter.

By no means all farmers converted their corn crops, or even part of them to silage. During the first half of the twentieth century, between

The wood-stave silo, fastened to the barn to prevent leaning, fermented and stored chopped green corn—stalks, ears, and all—for winter feeding of stock. The ensilage, shortened to silage, *had to be tramped by workers. It was one of the hardest, hottest jobs in the history of farming.*

5 and 10 percent of the country's production was chopped and stored in the big, tall structures that towered over the rural landscape. But all corn growers made use of dried fodder or stover in one way or another. The harvest activities described to this point stopped the developing of the grains of corn short of the mature, dry stage. Reaping of fully ripened ears greatly varied the tasks involved in the harvest.

Fodder pulling, a popular procedure of colonial and postcolonial times in both North and South, consisted of several stages. Just as the corn blades began to turn yellow, the settlers walked down the rows and "pulled fodder," taking all the leaves off each stalk below the ears. The workers wore gloves to prevent cutting their hands on the sharp edges of leaves. They bound bundles of these blades with other leaves and wedged them just above ground between several stalks that were conveniently close together in one hill. This prevented the leaves from blowing free and spoiling on the ground. After several days' curing, the leaves were hauled by sled or wagon to the barn loft or stacked in cone shapes around poles or small trees. The leaves at the tops of these stacks acted somewhat like thatched roofs and would shed moisture without serious damage to the fodder beneath. Many frontier farms, such as the Davis clearing in western Ohio, were ornamented in early fall with a number of these fodder stacks clustered around the log buildings. So long as the leaves were plucked before they became sun dried, cattle and horses preferred them to clover and other hay in the cold season. By pouring warm water over the cured corn leaves in winter, farmers made them more appetizing for cattle and at the same time increased the milk supply. However, fodder pulling hurt ear growth by as much as 10 percent and most settlers outside the South abandoned it by the early nineteenth century. Probably the greatest error of rural folk in harvesting maturing corn was their pulling or cutting of fodder too soon, before succulent leaves and stalks had contributed maximum development to the ears.

Some farmers as early as the 1700s advised cutting off the tassels and storing them for stock feed as soon as the grains began to harden; but A. Maynard, in a New England Almanac of 1849, wrote: "Let corn stalks stand till the tassels are entirely dead, so that the ears may grow and fill out." For those who had pulled fodder below the ears, a popular method was to wait a few days until the grains hardened, then cut the stalks just above the topmost ears and shock them in short stacks. By this means, called topping, they spread the work of harvesting fodder over a number of days, since the upper leaves remained green longer

Thomas K. Wharton's drawing of the Davis clearing near Piqua, Ohio, 1831, shows fodder stacks—corn leaves stripped from stalks and stacked around poles.

than those below the ears. After curing, these stacks would be hauled by sled or wagon to the barn or haymow for winter feed. Topping hindered the maturing of the ears even more than fodder-pulling and was practiced very little after 1825.

What was left in the field after fodder pulling and topping was an array of stalks stripped of everything but the ears. Some Indians and settlers broke the stalks over so that rain water would run off the husks; but this was not often necessary since the weight of the ears bent them down enough to shed water. In either case, the ears would "turn the weather" (shed moisture) and remain dry for weeks or months until the farmers had time between other chores to do the husking. Sometimes the ears would be shucked in the field, but probably more often harvest hands cut them with corn hooks or knives or broke them from the stalks and hauled all to the barn for the ever-popular husking bee. Those who husked early tested the corn in the field for ripeness by wringing the ear between their hands. If it squeaked as the shuck rubbed against the grains, it was not ripe enough to harvest.

A radically different and labor-saving harvest method was not to hand-harvest at all, but simply turn the hogs into a field and "hog down" the corn. Hogs were thorough in cleaning up grains from cob or dirt. Horses and cows could then forage on the remaining fodder. Throughout the pioneer period, hogging was a very widely used harvest method.

Cart or wagon harvesting was another of the many techniques. Some planters spaced their corn rows five feet or more apart and drove carts between them at harvest time. As a much more common variation on this method, farmers would plant their rows closer together and harvest by driving a wagon directly over one row, knocking the stalks down in the process. One or two huskers behind the wagon gleaned all ears from three rows with each trip across the field, the "down row" and one row on each side. Often a woman or child shucked the down row. Ed Donnell of Basora, Nebraska, wrote to his mother in 1885 that he wanted to "get married . . . before I commence shucking corn if I can to Aunt Jennie's oldest girl. She is 18 years old. She is a good girl and knows how to work. . . . She says if we marry right away she is going to do the work in the house and shuck the down row when I am gathering corn." A good team of animals was controlled entirely by a "giddap" and "whoa," leaving the hands of workers free to pick corn. Depending on time and weather, harvesters might husk the ears and toss them into the wagons, or simply throw them in with the shucks on to be husked later. Each wagon could be rigged with sideboards all around (to increase load carrying capacity) and a high bangboard on one side. The bangboard, when hit with a bang by an overthrown ear of corn, would cause the ear to drop into the wagon instead of going over the side and on to the ground.

The harvest method most damaging to the soil was turning horses and cattle into fields to eat leaves and stalks after wagon harvesting or hogging down. The hooves of the large livestock packed and temporarily ruined the soil. The resulting clods would only be broken by a winter's freezing and thawing.

At times it seemed that there were almost as many ways of harvesting corn as there were farmers. Yet one method, one system, one structure, which reportedly started in Hardy County, Virginia (later West Virginia) about 1780, moved across the Appalachians, through the South, and into the Old Northwest and Northeast like an invading army until it captured the nation's corn country. This was the corn shock, not to be confused with the fodder stack described earlier. The shock was a part of the rural scene from New England to Georgia as early as 1790, and by 1825 it had replaced the fodder pulling and topping almost completely and dominated the corn belts. Now almost as obsolete as the army mule, its place in lore and legend is just as firmly fixed. The corn cutters variously called the process shocking, stalking, stooking, and piking. Where once the overwhelming theme of the American landscape was timber, through the nineteenth and early twentieth century in the fall and early winter it was the corn shock.

Why the sweeping success of the corn shock? There were good reasons for its use, but there is no simple accounting for its dominance. Unlike fodder pulling and topping, it did not interfere with the maturing of the ears. The shock was highly functional as a way to clear the field and store the grain for future husking. It thus postponed much labor till slack times, enabling the farmer to concentrate on other chores. Perhaps the feature that set the shock above all the other harvesting practices was its art form. Probably no scene is more generally recognized as both esthetically beautiful and typically American than the field of corn shocks studded with golden pumpkins and crookneck squashes and stubble, framed with the flaming autumn foliage of hardwoods.

Cutting and shocking time varied with latitude, planting date, and the growing period of corn (usually from 90 to 120 days), but or-

Shocks on the Missouri corn farm where the author was born and reared.
The house, which is still standing, was built about 1820.
PHOTO BY THE REVEREND HOWARD D. HARDEMAN, 1948.

dinarily it fell between early September and mid-October when the brown-dry husks began loosening and curling on the ears and the grains were dry and firm. Farmers spaced their shocks so as to provide for the shortest practical distance that they would have to carry arm-loads of stalks from the point of cutting to the stack. An experienced farmer could gauge from a glance at the bulk of stalks about how many rows from the edge or corner of a field he should place the first shock. On rich soil with luxuriant growth, the shocks would be close together. Twelve to sixteen hills square or three to four hundred stalks was average for a shock.

Having established the approximate center of the cutting area for a shock, the farm hand would locate four sturdy hills of stalks adjacent to each other in a square, with two or three stalks in each hill. These would be bent over into two intersecting diagonals, then twisted, twined and tied about each other to form double arches joining three or four feet above the ground in the center of the square. The result was a firm brace called a horse, gallows, or gallus. An occasional corn cutter would drive a wood stake in the center, but a good shocker did not need this. A derrick-shaped ventilator made of poles and used as a shock center aided in curing the fodder, but was too expensive in time and work to be widely used. Now the center brace was ready for the bundles of cut corn to be stacked in its four right-angled recesses.

The heavy corn knives, much like machetes, had been sharpened, probably on a pedal grindstone, before leaving the barn for the day's work. This, of course did not replace that ever-present whetstone, carried in the back pocket to hone the knife edges occasionally, like stropping a razor between shaves.

Cutting corn was hot work. Making forty shocks in a day during a heat snap in September or a wave of Indian summer in early October was no small task for one person. There was little or no breeze among the tall stalks. It was therefore too hot to wear sleeves, but corn blades were truly blades, and their sharp edges could cut and bring the blood. At the very least a bare arm would be chafed and raw after toting a few bundles of stalks to the shock site. So the farmer would cut a sleeve from a worn out jacket or a pant leg from old overalls, and pin it to the shirt at the shoulder. A right-handed worker would pin the sleeve on the left arm and wear a glove or sock on the left hand. In this way there was adequate arm protection combined with cool clothing for the rest of the body.

Depending upon the height of the stalks, they would be cut six inches to a foot above ground and carried in bundles with the sleeved arm. The cutting stroke was a slight upswing. A downstroke would dull the knife blade from contact with the ground, and stopping each downswing would be much more tiring to hand and arm. Cutting about six hills to the armload, carrying each load with butt ends often dragging on the ground, and working around the double-arched "horse," the laborer would place bundles in the crotches, bracing them slightly outward at the bottom to avoid toppling and taking care to keep the shock symmetrical. After a few bundles were placed, they were usually tied to the brace for greater shock stability before the stacking was completed. Some corn cutters used a system called railroading. They tied gallows the length of the field; then cut one or two rows of corn all the way down the field, stacking each armload at the nearest gallows.

When the tepee-shaped, structured shock was rounded out to full size, perhaps five or six feet across at the base, the harvest hand tied it with a tight waistband of twisted cornstalks, with butt ends bent and thrust into the shock, a band of inner bark from basswood, wild grapevine, broomcorn, hemp, or in later years, binder twine, placing this tie

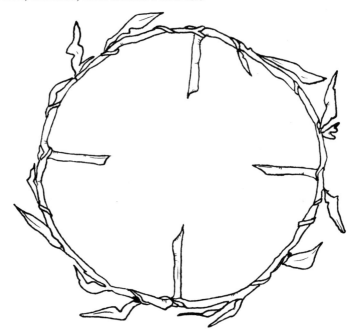

Top view of a tie band made of four twisted corn stalks used to bind the waist of the corn shock. The butt of each stalk was bent and thrust into the shock. Wild grape vine and binder twine were also used when available.

line about two-thirds of the way up from the ground. A firm tie was best achieved by use of a rope and notched wood block to cinch the shock very compactly, after which the twine, vine, or stalks could be tightly bound around the shock. Then the rope was removed for use on succeeding shocks. A sturdy horse, symmetrical stacking, and a tight tie were all of critical importance since a shock which began to twist would gradually settle to the ground in a corkscrew swirl, causing both grain and fodder to rot. Stalks shrank after cutting and the best farmers usually tightened the bands after about ten days to prevent this settling.

Spoilage was exactly what the corn shock was designed to prevent, and it did just that throughout the winter if necessary, except in the very deep snows of northern states. A New Englander wrote in 1790 that the shock "method is very favourable to drying the corn, if it needs it, as well as to defending it from rains," better, in fact, than "carrying it sooner to the stack or mow. There will be less danger of its taking damage by heating." The shock could be set up quickly, it presented no immediate storage problem in the limited barn, granary, or crib space, and it preserved both grain and fodder for months. Deer and rodents were greater threats to shocked corn than the weather. Storage in the shock enabled the farm family to turn to the myriad other, often more urgent, chores until there was free time to tackle the shocks piecemeal.

Among the more pressing tasks was often the preparing of ground and planting of winter wheat. Frequently settlers did this by picking the pumpkins, disk harrowing between and around corn shocks, and sowing wheat while the corn was still in the shocks. Other imperatives at the same season were digging potatoes and cutting a supply of winter wood.

Hidden away in each shock there was a gold mine of grain, and if ever there was hard work which was fun, it was shucking corn from the shock. On a given day the family work force, two or five or ten strong, would trudge to the field carrying water or cider jugs, biscuit or corn bread, salt pork, and apples for assuaging thirst and appetite, and likely as not a rifle-gun or shotgun. Country folk had a way of making do with what they had. If they were short a shucking peg or two, they could remedy this in a matter of minutes with a Barlow knife, a hardwood stick, and a strip of leather. Dogs and cats became a part of the act too, for this was more than a harvest—it was a chase.

At the shocks the teamwork began. The tie was removed from a shock and each person, man, woman, or child, took a bundle of stalks and laid them on the ground several yards away from the shock. Once this dismantling had stripped the shock away to half or a third of its original size, the excitement began. Dogs and cats were poised. In nearly every shock there were some tenants which had moved in to take advantage of free room and board. Mice and rats, rudely evicted from their corn stalk tepees, scurried out for the nearest neighboring shock or pile of fodder; and the smallness of the game scarcely diminished the thrill of the chase. Sometimes the fugitives were larger. An occasional possum ended its ungainly getaway effort by feigning death when overtaken and was picked up by its hairless tail and tucked away in a sack for future feast and fur. Minks, weasels, raccoons, and even yearling foxes and coyotes occasionally bolted for safety. Most dismaying, particularly to an inexperienced dog, was a skunk that had chosen a shock for shelter. It took weeks for the odor to wear off. As with porcupines, so with skunks—a waft to the wise dog was sufficient. Never again would it pursue the sluggish, bobbing, black and white bundle of fur.

The chase over, each person went to a pile of fodder, put on gloves and shucking peg, and began tearing off husks, snapping bared ears from stems, and throwing them into a common pile centrally located among the workers, usually in a basket or in the dry circle where the shock had been. When available, old jute bags, canvas, or other cloth would be placed on the ground where the ears would be thrown to keep them reasonably clean. Some farmers heaped the corn in two piles, one of prime ears and the other of nubbins (very small ears) and ears which were full length but had not been well pollinated and had developed few kernels. Youngsters engaged in some games and merriment and occasionally in petulant rivalries. One or another of them might wander over to the nearby persimmon grove and return with a cap full of the shriveled fruit, its puckery taste long gone, its skin turned bluish orange, and its flavor made sweet by the recent frost. Sly tosses of a few persimmon seeds, dirt clods, or corn kernels at co-workers were to be expected.

Shock after shock the routine was reenacted. There were occasional breaks from work when a dog wandered to the nearby woodlot and treed a squirrel. As the piles of grain and bundles of fodder grew high enough and the sun sank low enough in the west to tell them they had bitten off all they could chew for the day, the farmer or one of the older children would go harness the team and haul the wagon to the field. By collective effort the harvesters would load the wagon, tossing the good ears into the middle and front, nubbins to the rear. They loaded by hand, usually gloved hand, since the grains of dent corn in particular were rough. Homer Croy, an old corn farmer, said that those who neglected to wear gloves when shucking corn poured melted tallow on their cut and cracked hands at night to relieve the pain resulting from their carelessness. More modern cornfield toilers have been favored with a commercial corn husker's lotion.

After the harvesters had heaped the ears of grain onto the wagon bed, they placed the bundles of fodder crossways, lapping over the vehicle's sideboards on both sides and piled high. Because of the light weight of this stover, the load was not topheavy. It was lashed from front to back with ropes, and the younger children climbed the brace ladder at either end of the wagon and rode to the barn lot. There the fodder was removed and stacked where livestock could eat it as forage. Some fodder was used as litter or stacked around walls and stalls in the barn to help keep the animals warm. If it was raining or snowing or too late in the day to unload, the driver would leave the wagonload of grain under the wagon shed's protective roof. Unloading was much

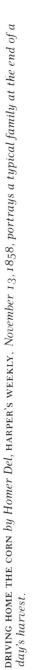

DRIVING HOME THE CORN *by Homer Del, HARPER'S WEEKLY, November 13, 1858, portrays a typical family at the end of a day's harvest.*

faster than loading since the flat wagon bed permitted the use of large scoop shovels for filling the corn cratch or crib. Such shovels for grain and manure were part of the equipment of every farm.

In many instances, the corn from the shock was pulled or cut by hand or by a corn hook and hauled to the barn unhusked. Sometimes the farmers did this to outrun a threatening storm, but more often it was to stock up for a husking bee.

Mechanization of the wheat harvest was nearly a hundred years ahead of corn largely because wheat could not be harvested as easily by hand and could not be left in the stack very long because of the danger of rain damage. Several wheat reapers, particularly Cyrus Mc-Cormick's, were in operation by the 1840s. The nature of wheat, with each grain encased by a hull, also made thresher development a necessity. These machines put the grain cradle and flail out of business. The principal parallel development in nineteenth-century corn harvesting was a horse-drawn corn binder of the 1870s and 1880s. Based on the wheat reaper idea, it cut corn stalks, tied them in bundles, first with wire, and after 1881 with binder twine (from a large roll carried on a spindle), and dropped them in the fields to be shocked by hand, as was the case with wheat stacking. But unlike the McCormick reaper, the corn binder did not bring a great change in corn cutting. It knocked too many ears off the stalks, and it was expensive. Small farmers, of which there were many, continued to wield the well-known corn knife. Another invention, the shocker, failed largely because its shocks twisted and settled badly.

A simpler device of the 1800s was the V-shaped cutting sled or wagon. This horse-drawn device had on each side a cutter which cut one row of corn. Cutters were sometimes made of old crosscut-saw blades. Two operators stood or sat back to back on the platform, each gathering the stalks from a row. Every few yards they would stop and shock the cuttings. Because it defied harvest mechanization for nearly a century longer than wheat, corn exerted a much greater force in pulling the pioneer family together as an interdependent economic unit.

Variations of the pioneer corn harvest were many, but with the principal exception of silage cutting, all involved human hands and that marvelously simple and effective Indian tool, the shucking peg. True, Obed Hussey invented a mechanical picker-sheller about 1838, but it was not successful. The huge, gasoline-powered pickers of the last several decades, harvesting as many as six rows at a time, have revolutionized all that, and have, in the process, contributed to technological unemployment and to the displacing of many millions of family members from American farms. About 4 percent of the nation's people now make their living primarily by farming, while less than 150 years ago that figure was nearly 90 percent. A New England sage could scarcely have foreseen such a change when he wrote in the fall of 1849, "Now husking commences, every ear must be taken in hand, one by one, stripped of husk and silk, and severed from the stock. This work will not soon be performed by machinery; the fingers—the wonderful contrivance of an Almighty hand—are indispensable in this operation."

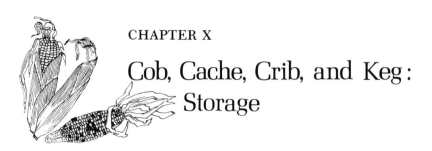

CHAPTER X

Cob, Cache, Crib, and Keg: Storage

Corn cribs are indispensable because this grain is preserved there longer than anywhere else.

CREVECOEUR, 1782

Some years ago an archaeological dig was being conducted in a region of central Mexico once inhabited by Aztec Indians. Picks, shovels, and other equipment and supplies had been hauled to the excavation site by mule. As the scientists worked with extreme care to unearth priceless artifacts, one of them chanced to look up and see a pack mule contentedly munching at something in an area of the diggings a short distance away. Sauntering over, the archaeologist found to his dismay that the mule was eating part of the future exhibit—corn, Aztec maize that was perhaps a thousand years old! There were probably several lessons to be learned from that experience. One of them, which is pertinent to this chapter, is very clear. Corn is a food which falls into the category of imperishables. In other words, it keeps. In fact, when properly dried and stored, it is one of the least perishable of the imperishables. One-thousand-year-old Peruvian corn still pops. Seven-hundred-year-old maize at Mesa Verde, New Mexico, is still edible and nutritious. Ancient Aztec corn appeals to a mule's palate. However grueling the work in producing corn, it was easier to store than most crops.

There is, however, another side to the corn coin. Few creatures—winged, quadruped, biped, horned, or even finned or bacterial—did not like it. Nevertheless, both Indians and newer arrivals to America met the storage problem with considerable success. With a few modifications, the Indian devices were adopted by early immigrants for corn storage.

Many wild creatures, including chipmunks, squirrels, woodpeckers, and beavers store food in time of surplus against a "rainy day." None, however, has been so successful as humankind. Between harvest and use, the Indians preserved their corn in two general ways: burial in carefully lined and concealed vaults, and storage in roofed structures above ground. There were so many variations of underground storage as to defy accurate description. Subsurface deposits were preferred for three reasons: they were relatively easy to dig, they preserved the corn adequately, and they could be well hidden.

Not always well enough hidden, however. On November 18, 1620, some Plymouth colonists at Cape Cod came upon a group of peculiar looking sand hills. Digging into these, they found Indian corn of various colors, stored in woven baskets. The intruders helped themselves. In later decades and centuries their descendants and successors discovered, copied, and sometimes stole from maize storage vaults of other Indians all across the continent. There were holes lined with tree bark, grass, stones, and clay. A favorite in the Northeast was a pit lined heavily with bark and relined inside with grass. Sometimes the native peoples stored mature and dry corn in such caverns. Other ears in these caches had been picked in the milky stage and roasted or smoked in the husk, then dried and stored, to be softened by recooking and eaten later. This was called parched corn. Husks left on the ears helped to protect the corn against insects.

Warriors often had numerous caches, each carefully camouflaged at the surface to foil the enemy, for a favorite method of combat was the scorched-earth policy, destroying the opponent's food supply. With many caches, some would probably go undetected and enable a tribe to survive. Several ingenious tribes of New Mexico stored corn in stone rooms constructed beneath overhanging cliffs and accessible only by removable ladders from below. The Incas of South America had rectangular clay and grass boxes with vertical slotted windows so that they could see when the maize supply was running low. Pottery vessels, baskets, and sections of hollow trees provided added storage facilities. Among the Mandan, Hidatsa, and Arikara of the upper Missouri River region, some of the corn storage pits were so deep that ladders or scaffolds were used to enter and leave. Such excavations, of course, had to be carefully concealed from human and beast, and it was not unknown for unsuspecting wanderers to fall in. Caves, when accessible and dry, also served well for grain storage.

Rot, rodents, mold, and flooding sometimes spoiled Indian corn caches, but all in all, the native peoples were quite successful with their invisible underground storage caverns and were able to leave them unattended for long periods while away on hunting or war trails.

Except for silo pits and cellars under their dwellings, however, settlers did not copy the Indians' underground storage systems for corn.

More to the pioneer farmers' liking were several types of visible corn storage structures devised by the Indians. Either in mature dry stage or parched, ears of various colors were hung by their braided husks or in baskets or pottery around the interior walls of the dwellings. Suspension kept them free of damage from mice, rats, and squirrels; and birds did not often venture into the structures. The festooned, varicolored bundles of corn were things of beauty, like pictures or works of sculpture or some artist's design of lighting fixtures. Whether it was the inherent beauty, the practicality, or both that appealed to the small farmer settlers, they adopted and never abandoned this Indian corn storage technique, and it has remained to this day as home decorative art.

Small farmers or new settlers in an area often relied on storage in their cabins until they had time to build separate drying and storage structures. The attic above the kitchen area, because it was the most frequently heated part of the dwelling, was a favorite storage space for small quantities of corn.

Another Indian corn-storage creation, which immigrants copied and maintained as their most important system, was the cratch or crib. Hernando de Soto's company of explorers found it in use in the Mississippi Valley during the 1540s. It was a little house with bark or thatched roof, sides of slightly separated slats, and a cane floor, and it stood on four posts. While not rodent proof, it resisted the little varmints somewhat. The crib at once kept the ears of corn aired and shielded from direct downpours. There was far less problem from mold and mildew than in storage pits.

Such four-poster Indian cribs existed in the Northeast and in other wooded parts of America as well. Some sources attribute the crib design to Euro-Americans. While they had such structures for crops (such as hay) and certainly improved on the Indian pattern of corn cribs, the origin of this design for corn storage was Indian.

Indeed the design underwent many changes at the hands of settlers, although 350 years after the arrival of de Soto sixteenth-century tradition was being built into the lines of the new cribs, as surely and as clearly visible as saw cuts on the timbers. Pehr Kalm describes the cribs in 1750 as about eighteen feet long, two to three feet in width, and of whatever height desired for the amount of corn to be stored. Sides and

The typical corn crib, or cratch, had slatted sides to ventilate corn ears. Inverted pie tins kept rodents out.

bottoms were slatted for admitting air, and the roofs were broad—not unlike Swedish haysheds, Kalm observed. Typically, Americans designed the cribs with inward sloping sides, narrower at the bottom, sometimes with hinged storm doors or flaps to keep out swirling snow and rain. The foundation posts of the cribs were models of evolutionary experimentation to keep out rodents—smooth, slick-peeled logs; hollow clay tiles; metal sheathing; squared posts with glass panes overlapped in shingle fashion, anything to thwart little climbing claws and gnawing jaws from reaching the precious winter hoard of grain. Eventually, inverted shallow tin pans (often pie tins) proved to be effective and inexpensive rodent stoppers.

Corn cribs were by no means standardized in design. The V shapes were most popular, but there were simple log sheds without chinking (as chinking would have restricted air circulation); and there were drive-through models where wagonloads of corn could be scooped into covered, slotted bins on both sides at once. Too, there were the simple gable-roofed, barnlike buildings that served well as cribs with room to scatter the ears over the floor. A half wall down the middle of these permitted the storing of grain-laden ears on one side and throwing of corn cobs on the other. Settlers always saved cobs for a myriad of uses.

On many farms there were several small cribs to provide ample storage space combined with optimum airing. Ears in the middle of a very large, filled crib would not have aired properly. A width of nine or ten feet was considered the maximum for proper airing. One feature of the cribs was that they were usually far enough from each other and from barns, wagon sheds, hay ricks, and other structures to permit air to circulate freely.

The shock and silo were corn storage facilities that were more than highly significant. They were little short of sacred American institutions. Jute, duck, canvas, and deerskin sacks filled with ears were tied to rafters of buildings. A series of ears tied to a single length of twine also served well. And the soft, pithy cores of cobs could be pressed onto nails so that ears of corn would stand out on rafters, beams, or poles like limbs on a "corn tree" (see page 72). Similarly the ears were speared on specially designed, suspended wire racks. Corn husks and cobs themselves must be viewed as important corn storage accessories, the former as hangers of ears or protectors to better retain flavor and keep out insects, and the latter as holders that permitted proper aeration until shelling time.

Shelled and ground corn presented storage problems of an entirely different type. Lacking the plastics and metals of modern times, the early farmers had to secure wooden containers for large quantities of small granules. They had some success in storing dry kernels in log buildings made almost airtight by clay chinking to keep out weevils, although at the price of reduced air circulation. Sacks were useful, but vulnerable to rodents' sharp teeth. The wood containers were also vulnerable, but oak kegs and barrels were far superior to jute or cotton sacks. Thus came about the "marriage" of the cooper with the storage and shipping of grain and meal. The skilled cooper, in an amazing feat of hand-tool ingenuity, could make barrels, hogsheads, and kegs with such tight-fitting staves that they would not only hold cornmeal, they would even contain water. Those who made barrels for water, whiskey, whale oil, and the like were called tight or wet coopers, while slack or dry coopers fashioned the barrels for grain and meal. Cooperage was indispensable to the storage of corn.

Nature was its own cooper as well. Many a settler who could not afford the price or the time for barrels to store shelled or ground grain simply walked along the stream bottom land, or the slightly higher second bottom, and rapped on the bases of the big, gray-and-white-barked sycamore trunks. Nature had a special bent for leaving many of these trees hollow. Some of these were so large that small families slept in them until cabins could be built. A few taps with a stick or an axe would tell which trees to cut. With a helpmate to handle one end of a two-handled saw—a tool brought from Europe by the first colonists—the farmer would soon have one or perhaps several barrels or kegs cut. After a minimum of chipping or sometimes burning of the soft, loose inner rot, and nailing a bottom slab on each wooden cylinder, the pioneer had some rustic but serviceable storage containers which would last for years. Containers of this type were called gums, perhaps from the term *bee gum*, a hollow tree taken over by a swarm of honey bees.

Modern America has largely solved its grain container problem with cylindrical tanks of corrugated metal. These are rain and rodent proof and resistant to weevils. Esthetically it might be argued that they are less pleasing to the eye than shocks, cribs, and barrels.

Railroad terminals and shipping ports created a need for much larger storage space for corn and other products since they were holding points for wide geographical areas. Huge grain elevators wrought new changes in the skyline of America, as did the railcars and locomotives that hauled the grain from farm to market. This commercial phase, of course, called for much shelling of corn since cobs are about equal in volume to their kernels, and it would have been less economical to ship the cobs along with the grain in railroad cars and cargo boats. It also necessitated artificially drying the fully matured grain from its normal fourteen or fifteen percent moisture down to twelve percent or less. Before such dehydrating was done, it was not unusual for steaming hot water to gush out of corn bins after shipment. Indians and pioneers alike had preceded modern drying techniques with their mats in the sunlight and fires built under corn cribs. But the mass shipments called for dehydrating great volumes of grain just as they made necessary the shelling of corn in quantities impossible to shell by methods available to the settlers who opened the farming frontier.

CHAPTER XI

Shelling and Grinding

Your boys had better be shelling corn than playing snowball.

ROBERT THOMAS, Massachusetts, February 1802.

It was November, 1800, on the Tennessee frontier. A novice corn-farming family, of which there were many, hitched a team to a creaking wagon, loaded their barrels of dried corn on the cob, and drove at the slow gait of the lumbering oxen to a neighborhood grist mill. The trip along the timber-lined road was not difficult. Of all the country roads, those to the mill were the most traveled and the best kept. Arriving at the mill house with its pond, its headrace, and high, trestled sluice, and its big, carnivallike wheel, the visitors were greeted with distressing news. Yes, the miller could grind their grain with due attention to their requests for coarseness or fineness, but there were no provisions for shelling corn at the mill. His explanation was simple: there was no mechanical way to shell corn, and he had not the labor to do it by hand. The farmers must shell it themselves.

Such a scene, reenacted more than a few times in the westward motion of the great farming frontier, was evidence of the fact that grinding, despite its seemingly complex problems, was many centuries ahead of the technology of removing grains from cobs. Most corn shelling would be done by hand for another forty years. The word *shell* has several definitions. One of the simplest is the removal of corn grains from the cob, which was sometimes done by use of a seashell. (Perhaps this was the origin of the term.) But the process for centuries—millennia in fact—was by no means simple if there was any large quantity to be shelled. The point of each grain is imbedded firmly in its cob. One grain is not difficult to remove, but extracting one at a time from a hundred ears, each with five hundred to a thousand grains, would take an eternity. To make matters worse, the ker-

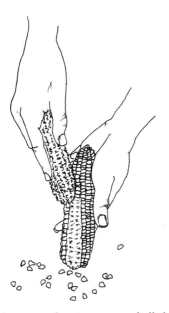

The Indians taught pioneers to shell their corn by knocking out two rows of kernels with a corncob and twisting off the remainder with a wringing motion of the hands.

nels stand solidly together, like a Greek military phalanx, reenforcing each other. Chewing or cutting the milky grains off the cob, cooked or raw, was simple enough, but the processing of meal or flour required first drying the grains on the cobs and afterward shelling.

As in many aspects of agriculture, little progress was made in shelling corn from the dawn of plant domestication until near the middle of the nineteenth century. The job was trying enough that Indians and pioneers alike usually shelled only what they needed for the time being. Some of the techniques, though elementary, were practical. One method involved placing the ears on hard ground, preferably on a blanket, and beating them with a club until enough grains and rows were loosened so that the remainder could be twisted off with a wringing motion using both hands. A second and similar practice was to roll the ears in a hammock of fibers, suspend it over a blanket, and beat the roll with clubs, causing loosened grain to fall through the coarse weave to the blanket below. These Indian methods, including the use of the aforementioned seashells, were copied by early settlers.

Another shelling technique appropriated from the natives became very popular with pioneers. The worker held one ear at a downward angle in one hand, and with a corn cob, stick, or bone grasped in the other hand, knocked off two adjoining rows of grains. This would break the phalanx and the other grains could be quickly twisted off sideways by the above-mentioned wringing, twisting motion of both hands. Gloves were useful in this operation to avoid blisters. An Indian device very little used by white and black pioneers was the cob disc. Many cobs were tied firmly together in a disc or cylinder eight or ten inches across. The performer of this chore then briskly rubbed each ear of corn over the ends of the cobs until enough grains were removed and loosened to twist off the remainder.

Pioneers soon went beyond the copying stage and devised shelling methods of their own. Livestock, particularly horses, were driven or led repeatedly over a bed of unshucked corn on the barn floor, and grains, loosened by the animals' hoofs were twisted off by hand.

This was a haphazard method, however. Also less than satisfactory, although much faster than shelling by hand, was the flail, or swingle, a two-foot piece of wood tied by a thong to a six-foot wood handle, and used to beat ears of corn on the barn floor until the grains were loosened. Farmers of the Northeast in the mid-1700s were shelling corn by strapping a sturdy piece of iron such as a large hinge, a shovel, or the handle of a big frying pan, to a tub and dragging the ear across the edge of the iron. This broke the grains loose from the cob and they fell into the tub. By the end of the century, a slight variation on this method had become very widely, almost universally used in the seaboard states. It involved adapting what was popularly called Uncle George's toasting fork, a triedged bayonet from a Continental Army rifle. (*Uncle George* was a reference to Washington, the commander of the Revolutionary Army.) The bayonet point was driven into the inside of a large wooden tub several inches down from the top. A piece of hardwood was thrust into the bayonet socket and fastened to the opposite edge of the tub. Each ear of corn was rasped downward across the upper edge of the bayonet until enough grains were removed to twist off the rest. The job was done indoors, usually at a time when bad weather interfered with outside chores. A good worker could shell twenty bushels (about eleven hundred pounds) of grain in a day by this method.

Such were the primitive shelling methods until the first half of the nineteenth century when a rash of new sheller designs came into use. The earliest of these, invented about 1800, were iron-toothed wood cylinders cranked over toothed wood staves arranged in semicircular form close beneath the cylinders. Attendants fed ears of corn between the cylinders and the staves, then cranked the cylinders, causing the teeth to tear the grains loose. These kernels in turn fell between the

"Uncle George's toasting fork," a musket bayonet of Washington's era, was driven into the side of a wooden tub and used to hand-shell corn. Pehr Kalm described a similar method in use in the 1740s.

staves and into basins below. Thomas Jefferson left a sketch and written description of one such machine, made by Paul Pillsbury of Newburyport, Massachusetts. This sheller had an oak cylinder two feet in diameter and seventeen inches long, with numerous small teeth projecting about three-eighths of an inch from the cylinder. A similar model of the same era had spiral metal cutters running the length of the cylinder. This principle was applied to a belt-driven, horse-powered or steam-operated machine, Smith's Corn Sheller and Separator, which was invented in the mid-1840s. The cylinder was about six feet long and fourteen inches in diameter and was encased in a metal tube. The machine cost fifty dollars and shelled as much as three thousand bushels of ears (fifteen hundred bushels of grain) per day.

But the average farmer had no need for the quantity shelling made possible by Smith's machine. He needed a small-scale, cheap, reliable design. By the late 1820s, several inventors made shellers that reversed the Indian principle of the cob disc and set the trend for sheller

Burral's corn sheller, developed in the 1840s, used a spur wheel in the central housing to tear grains efficiently from the cob. After slight modification of this basic design, such inexpensive devices were soon to be found on practically every farm.

designs until the day of the modern combine harvester. By turning a hand crank, the operator rotated an iron-toothed wheel which simultaneously pulled and rotated each ear, and tore the grains off as it was fed endways into a funnellike hole from a hopper. The grains, as in more primitive methods, dropped into a basin below. Except for the homemade hominy block (discussed later), the small corn sheller, cranked by one operator, was the first genuine corn machine to become standard equipment on most farms. By the 1840s, Burrall's, Clinton's, and Taylor's devices were on the market, and by mid-century nearly every corn producing family had "shelled out" the ten-to-twenty-five-dollar price for a sheller. Conservative agriculture surrendered more quickly to the corn sheller than to any other innovation. In fact, the sweeping of the corn country by the hand-cranked sheller probably represents the most rapid acceptance of a major technological advance in the history of agriculture.

In time, designers modified the shelling machine to toss the cobs out into a different container. Soon they greatly enlarged the hopper and added a feeder hole to shell two ears at a time. Some small shellers were bolted to tables. Larger ones stood on the floor. A heavy flywheel was added to ease the human burden on the larger models. Then came the use of other energy—water, horse treads, steam, and eventually gas and electricity—to run the sheller. In the course of time, fanning mills separated the few particles of chaff from the grain. Now, combine harvesters pick, husk, shell, and fan the corn in one operation, dropping the grains into a truck driven alongside. And those picturesque shellers of yesteryear are today's museum pieces. Shelling and grinding were the most intricate of the corn-related tasks. Necessity had become the parent of both invention and production.

Shelling must precede grinding, but efficient mechanical corn

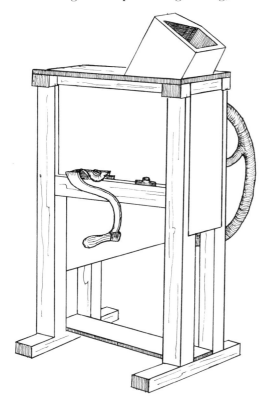

This corn sheller of the mid–nineteenth to early twentieth century used the same spur-wheel principle as Burral's sheller, but a heavier flywheel and spurs on both sides of the enclosed shelling wheel sped up the process.

shellers were developed long after good grinders. There were several reasons for this. First, the ancient and advanced civilizations of Europe and Asia had thousands of years' experience in devising machinery that would crush wheat, rye, barley, and other grains. Such grinders merely took in another grain—Indian corn—and handled it easily since maize was much less likely to gum up the millstones than were the smaller grains. Second, the process of random grinding of grains between two hard surfaces posed less of a technical problem than removal of tightly packed grains firmly attached to a woody core. Finally, the old ways worked despite their slowness, and there was a ready labor force, composed primarily of women and children, to handle tedious jobs such as shelling corn in Indian and pioneer America.

For that matter, there were probably many more "Ginnys" than "Jimmys" who cracked corn on individual American farms of the early era, despite the lines of the well-known folk song. Among Indians, whites, and blacks, women usually operated the people-powered

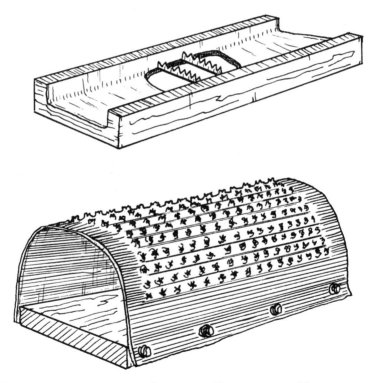

Gritters, or graters, were used to grate milky green corn. The name may have been related to hominy grits.

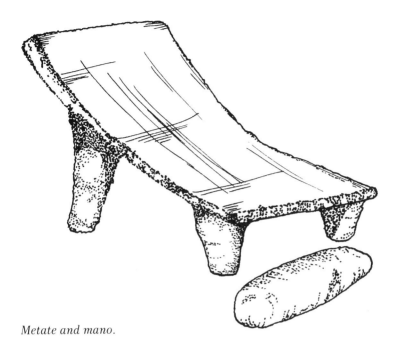

Metate and mano.

cracking and grinding devices. Ground corn was both a necessity to convert hard, flinty kernels to a more digestible form, and a luxury because it lessened the monotony of eating the same product day in and day out by varying the ways of preparation. Different consistencies and tastes were achieved by grinding corn coarsely or finely, as well as by drying, parching, soaking, boiling, or treating with alkali or lime.

Tracing the origins of mechanisms for grinding grain is like trying to identify the inventor of the handkerchief or dishcloth, or the first crooner of a folk song. It is clear, however, that pioneer Americans made use of the whole range of grain grinders known to humankind. Most of these had been employed in Europe, and the ideas, if not the machinery, were brought by immigrants to the New World.

The simplest of grinding implements was the grater. Indians used stone knives or the sharp edges of large stones for grating soft corn from the cob. Metal strips used by the settlers were an improvement over stone. These were called tin gritters and were fastened to boards for stability during the grating operation.

Among grinding implements, Indian metates and manos and European mortars and pestles were the most primitive. They were the same in principle although somewhat different in appearance. Each had been widely known for thousands of years. Eastern and western hemispheres learned from each other and at times pirated designs from one another. *Metate* was the Aztec word for a device known by most tribes of the Americas. The Indian woman crushed the grain between a hand-held stone and a flat or concave bed stone. A metate was sometimes small and mobile and sometimes a depression in a massive rock outcropping. Usually, according to descriptions of European visitors, it was dirty from decayed particles of grain, meat, or nuts.

Most metates were made of granite, although basalt, limestone, and sandstone were also used. One serious problem, worse with sedimentary rocks such as sandstone, was that granules of rock wore loose and were chewed with the corn. Despite some success in washing or winnowing to remove the tiny stone particles, they caused serious tooth wear among the tribespeople. Metates, although widespread, were most common in the American Southwest, and non-Indian and mestizo settlers in that area adopted them readily.

Two variations on the metate design were widely used in the eastern woodland. One was the mortar and pestle, which had evolved from the metate-type saddle stone developed in the Old World of prehistoric times. Immigrants introduced it from Europe and relied upon it to some extent in America for home corn grinding. The mortar was a stone or metal vase and the pestle a grinding stick, metal rod, or piece of stone roughly resembling a stubby baseball bat. Another variation was a large wooden mortar, which was used by Indians of the eastern woodland and copied by newcomers from across the seas. It functioned in the same way as the metate and the mortar and pestle.

By far the most important and widely employed home grain grinder throughout the timbered regions of Anglo-America's corn country was the hominy block, or samp mill. Both these names were derived from Indian words. *Hominy* was an anglicized Algonquian term for hulled corn (ground or whole). *Samp* was from *saump*, a Narraganset expression for coarsely ground corn or cornmeal porridge. Settlers also called this simple but ingenious device a sweep-and-mortar mill and a corn cracker (although several types of grain crushers were known as crackers). The word *cracker*, as applied to anyone from the poor white class of southern hilly regions, seems to be traceable to the heavy reliance on corn cracked in such mechanisms.

In form as in name, the hominy block was similar to the Indians' wooden basin and pestle. But its design was an improvement and di-

Nearly every pioneer farm had one or two hominy blocks, or samp mills, like this one for cracking corn.

rections for making it were simple. Near the cabin, cut off a hardwood tree (preferably oak or hickory) three or four feet above ground and hollow out the top. From a springy limb of another tree extending over this stump, tie a pestle or block of wood by a strong line. To operate, grasp one or two limb stubs, which have been left on this hanging block for handles, and plunge the suspended piece repeatedly into the cavity of the stump top to crush the corn kernels placed there.

The tree limb, or sweep, acting somewhat like a heavy fishing pole, lifted the pestle or hammer between each downstroke, thereby greatly reducing the human effort. In this manner, dried corn grains, which had first been soaked in water, were pounded until the tough skins or pericarps were loosened and could be washed free from the cracked particles called hominy grits. Originally that term probably referred to *grates*, coarse particles of corn kernels grated on rough, perforated tin grates.

Many hominy blocks were enclosed within cabins, barns, or other shelters to provide all-weather use. Lifter limbs were cut from trees and braced under roofs of the buildings. When choosing the site for their cabin, a frontiering family nearly always considered its location in relationship to trees which could be cut to make hominy blocks. These mills were as much a part of the pioneer American scene as fireplaces and rain barrels. By the early eighteenth century, every cabin and clearing had one or two of them. Sailors could sometimes tell when they were nearing the East Coast in a fog because of the "thump, thump, thump" of samp mills. And pioneers used these thudding contraptions like signal drums to communicate with each other. Southern planters, who forbade their slaves to have drums for fear of insurrection, were at times apprehensive that the blacks were beating signals among themselves with the ever-present hominy blocks.

A lesser-used cracker, which operated on much the same principle as the hominy block, was the plumping mill, such as was built by the Blake family, small farmers of Dairytown, Connecticut, in 1805. Its power came from the weight of water tumbling into an open box on a lever-and-fulcrum mechanism. When filled by water from the end of a spout or trough, the box would tip the lever downward, lifting a wood hammer on the other end of the seesaw lever. Water then spilled out the low back end of the box, lightening it and allowing the hammer to fall on grain which had been placed in a hole atop a stump. It was less popular than the samp mill, largely because it lacked the precise directional control of the falling hammer.

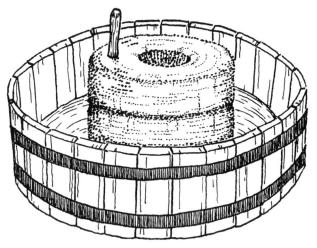

The operator of the quern, a small-scale, single-family device for grinding grain, poured the corn through the cone-shaped axle hole at the top. The handle was used to rotate the capstone on the stationary nether stone, and the meal worked out between the stones to fall into the tub.

Nevertheless, the plumping mill principle was applied to a very practical use. Farmers of the early nineteenth century were beginning to recognize the value of crushing the ears of corn, both cobs and grain, as livestock feed. Cobs had about one-eighth the food value of kernels and were useful for roughage in stock feed. Furthermore, crushing cobs and kernels in the same operation eliminated the time-consuming job of shelling. In the summer of 1818 a North Carolina inventor set up a heavy, iron-plated plumping pestle that fell at the rate of two strokes per minute. It was left to work all day on one load of ears in the mortar. By evening the grain and cobs were "reduced to a very fine meal," and without any attention required. In the 1840s other cob and corn crushers were developed, notably by John Pitts of Rochester and that prolific inventor Obed Hussey of Baltimore. These were horse-powered machines with cutting blades much like wood plane blades. Power grinders of the late nineteenth and twentieth centuries would again use metal plates and burrs, but throughout most of the pioneer period, settlers relied on stone.

The quern (pronounced *kwern*) was a small, stone-burred grain grinder apparently invented in ancient Rome. (Icelanders called it a kvern.) It used a principle probably new to the American continent— a stationary nether or bottom stone, sometimes called a bedder, and a horizontally rotated cap stone, or tedder. This upper stone was turned around an axle by use of an off-center hole and a twirling stick manipulated by hand. The user poured corn through a hole in the cap stone and pulverized it between the two flat surfaces of stone. Because it was small and easily operated, the quern was popular in all areas as a home grain grinding device. On the frontier it was the first rotation stone grinder in use, as William Byrd observed on his surveying expedition of 1728.

Revolving mill stones triumphed over metates as convincingly as guns over spears. The horizontally rotated stones were by no means limited to single-family, quern-type operations. This principle was applied on a small business commercial scale throughout the American corn and wheat country, beginning as early as the 1620s. Energy was supplied for many mills by livestock, and occasionally by humans and wind towers, but most often by water. For animate energy, a crossbeam was fastened to the cap stone by metal ring pins and the animals walked around and around in a circular motion on a tow path, using leverage to rotate the stone. Settlers from England referred to this as a turnstile. The Spanish called it an arrastra. It was common at the California missions, where nursing mothers were often assigned to grind the corn into meal.

Water power was much more complicated, since it usually required mechanical transfer of force from a vertical water wheel to horizontal millstones. The design had been worked out by ancient Greeks and improved upon by the Romans. In America, the power transmission was accomplished with the use of oak gear teeth fixed in holes on the perimeters of large wooden wheels with timber shafts. As with all types of revolving stone grinders, the miller poured grain through a hole beside the vertical axle at the top of the cap stone, and as the corn was ground, centrifugal force and burrs, outward-radiating grooves and ridges cut on the cap stone grinding surface with a hardened steel mill pick, worked the meal outward until it fell into a gutter or groove. This channel passed completely around the nether or bed stone and had a spout like a pitcher at one point. Under the spout, the operator placed a barrel, bucket, or bag to catch the meal. Millwrights tried direct-drive, vertically positioned millstones, for example at El Molino Viejo (The Old Mill) near Los Angeles, but gravity carried the meal through the grinding cycle too rapidly, and regrinding was thus more often required.

For the amount of water used, the most efficient among the four basic types of water wheels was the "overshot," since it was driven by both the striking force of water flow where the flume emptied at the top of the wheel, and gravity power from the weight of water in its down-turning buckets. The "breast mill" was probably second, having water fed to the buckets at axle height instead of the top of the arc. Overshot and breast mills required controlled falling water. Since natural sites were not always located where mills were needed, some grist mill operators frequently built dams on streams to create ponds. Others took over the engineering work of colonies of beavers, channeling waters from these ponds through millraces and wood troughs or flumes into the bucket blades of the mill wheels.

A third design, the undershot or flutterwheel, received the force of stream flow only on its blades at the bottom, and compensated for the lighter rotation thrust per unit of blade area by having wider paddle blades. It stood over a broad stream of water rather than under a narrow trough or flume. An undershot mill also required less complicated engineering of the water flow. The tub mill, a fourth type, differed from the other three in that it consisted of a horizontally fixed water

Draft animals were hitched to the wooden bar of this mill. Pulling in circles, they would rotate the capstone. Whole grain was poured through the funnel, and meal worked out between the stones. Although used throughout the country, this design was most common at missions and pueblos from New Mexico to California.

wheel with slanting, fanlike propeller blades cut in it. Water flowed down through the propeller, which turned a direct-drive vertical shaft, with the cap stone rotated by the shaft above the power wheel. In colonial Virginia, the first water-powered mill was built in 1621. By 1649 that colony had five water-driven mills.

By the mid-nineteenth century, there were steam-driven grain mills in most sections of the country, although they were not numerous. Santa Fe, New Mexico, for example, had a single steam-powered mill in the 1850s.

As in so many other aspects of the corn business in New Mexico, mills and milling differed somewhat from the machinery and practices of the Anglo-American culture to the east, particularly before the occupation of the province of northern Mexico by the United States. The people of New Mexico grew corn almost entirely for human consumption. There were numerous mills throughout the settled areas. Most of these were of the direct-drive type, arrastras powered by horses and mules, and tub mills driven by water wherever it was accessible. They were usually small units with very limited capacity compared with the big overshot types. After the arrival of American military forces in the late 1840s, a number of New Mexico's residents took advantage of business opportunities by contracting to supply the army with meal and flour. Among these were Antonio José Otero, Simeon Hart, and Cerán St. Vrain. St. Vrain went to Westport, Missouri, in 1850 and bought the parts for several mills, setting up at least three in New Mexico during the 1850s. These were of much greater capacity than the older mills of that territory.

THE OLD MILL, *from* HARPER'S MONTHLY MAGAZINE, *1857, depicts an efficient overshot-wheel type.*

In early California, mortars, pestles, and arrastras were used to pulverize maize and other grains. Mission fathers later built water-powered mills at San Gabriel, Santa Cruz, San José, and San Luis Obispo. Numerous mills were built in the American era.

A maker of the whirling wooden mill wheels was called a mill-wright. Since there was not likely to be enough business in one locality to keep a millwright employed, the wheelwright, or wagon maker, was often contracted to make a mill wheel. James W. Marshall, discoverer of California gold, was both wheelwright and millwright.

Mill outputs, of course, varied. Because most local farmers' milling needs at any one time were usually small, the big wheels seldom operated to capacity. The *American Farmer* of February, 1832, stated that ten hogsheads of eight hundred pounds each (nearly sixteen bushels per hogshead) was a moderate day's output for a stone grinding mill with a "good pair of burrs."

Where could a prospective mill owner get such a pair of burrs, or millstones? For hundreds of years, France produced the best stones from quartzite deposits of the north-central region, and French burrs are still used for wet milling in the United States. Countless trading ships came to the American colonies from Europe with millstones in their holds. Granites from Vermont, New York, Pennsylvania, and most states of the Appalachian chain as well as from the Ozark Highland were widely used, particularly for grinding corn to feed livestock and poultry. Many millers had several grades of stones for different fineness of flour and meal, although tightening or loosening the cap stone helped to regulate fineness. Millstones usually ranged from three to seven feet in diameter and up to a foot in thickness. A small stone was made of a single slab, while a large one was often fashioned from a number of pieces mortised together and firmly held with an iron rim. This metal band, put on when red hot, contracted upon cooling, like the iron rim of a wagon wheel.

While shaping millstones was a long-established specialty, in the absence of an artisan of this trade many a stone mason was called upon to do the job. Before he left for the California gold mines, Robert Brownlee, an Arkansas building-stone mason and a corn farmer in his own right, met the needs of local millers by cutting and burring millstones.

Until about the mid–nineteenth century, there were several standard characteristics of the water-powered, stone-grinding mill businesses. Most employed the efficient overshot wheels, and all of those that produced meal were dry milling units. Wet milling emerged later, was typified by much larger capitalization, and was related to production of such commodities as corn syrup, oil, and starch. Although some, particularly in New England, were community enterprises and others were partnerships, most of the stone-burr dry mills were run by single-family operators, who were so indispensable that they were usually exempted from military service. Almost invariably the families bringing corn to be ground had to shell it themselves, since there were no shelling facilities at the mills. The customary practice for millers was to keep a toll, or share, of the ground meal as pay for the grinding services. This fee was often set by law and was usually one-twelfth or one-sixteenth of the amount ground for a customer. The mill owner sold most of his share to settlers who were short of grain. In quantity milling, deception was easy and some millers had to be carefully watched to see that they did not "weigh short," or keep too much toll. For some reason, the legend developed that the miller was not to be trusted, just as a companion legend held that the miller's daughter was always beautiful and pure.

Saw mills, grain mills, and whiskey distilleries, or "stills," were often operated jointly, since the same wheel and power could be used to saw and grind. Ground corn, the chief product of many mills, was also the principal material for making whiskey. Finally, and of utmost importance to good business, all mill owners and operators had to keep their mouths discreetly shut since, from the nature of materials brought to be ground, they could tell who the (other) neighborhood moonshiners were. It would have been bad business for neighbors to "tell" on each other once the moonshining business became illegal.

Wind-powered mills, like the water-driven models, had a rich history, dating from twelfth-century Europe, to Governor George Yeardley's corn mill built at Jamestown, Virginia Colony, in 1620 and to William Cradock's grinder at Watertown, Massachusetts, about 1630. But in the heavily wooded lands east of the Mississippi, timber blocked the breeze. Millers did not widely rely upon wind power until farming reached the treeless Great Plains in the post–Civil War era. The stone-burred Dutch Mill, built near Wamego, Kansas in 1879 and since restored, was an impressive but nontypical example. Most western windmills were used for pumping water rather than for grinding, although merchants, particularly mail-order houses, sold a number of steel-burr grinding windmill models during the 1890s and the early years of the twentieth century.

That astute observer of human affairs, Benjamin Franklin, noted that grist mills were the connecting links between agriculture and industry. But they did not live by bread alone. Left entirely to flour and meal, the wheels would have been idle much of the year since a disproportionate part of grinding was done between November or December and April or May, the season following the harvesting and proper drying of corn. During slack days and months, millers connected oak gear teeth to linkages which drove saws, broke hemp, fulled cloth, or pressed cider. The grain mill wheel of Isaac and Rachel Blake in Connecticut operated a saw and the bellows of a forge. A grist mill at Plymouth, Massachusetts, became Terry's clock works and later Seth Thomas' clock factory. The half century of trade disruption between England and America from 1763 to Waterloo and the end of the Napoleonic wars greatly stimulated industry, particularly in New England and the middle states. Numerous grain mill sites became factories.

Franklin might have noted, too, that mills very often grew up to become villages and towns. From millrace to main street was but a few steps. From Upper Mills and Mountville, New York, to Smyrna, Tennessee, and Bartlesville, Oklahoma, the cluster of town buildings followed construction of grist mills. Southern plantation mills were less likely to become towns than milling centers of the North and West. If a town lacked a miller, its citizens would probably recruit one, as did the settlers at Eastchester, New York Colony, in the late seventeenth century. And there were times when another mill was too much. In 1804, the Robertson, Brooks, and Booker families, all millowners, urged the Williamson County, Tennessee, Court of Pleas and Quarter Sessions to refuse a permit to another prospective miller. To no avail. The court opened the door to new competition by granting the permit to Thomas Hardeman, who ground grain, made whiskey, and sawed barrel staves at his mill on Big Harpeth River.

Mill owners were a diverse lot. Sometimes they were or would become prominent political figures like ex-President Thomas Jefferson, William Byrd, or George Rogers Clark, of Virginia (although Clark's mill was in Indiana); Newton Cannon with his horse-powered mill in Tennessee; or Peter Burnett, lawyer and small farmer of western Missouri and destined to become first governor of the State of California. More often, they were the "little" people, the Blakes on a tiny Connecticut farm; the Carrs of Indian Territory; or the John Plummers of Jasper County, Missouri, who prospered grinding corn for the Indian trade, and the countless frontier farmers who saw opportunity in the needs of their neighbors and in the swirling waters of wooded streams. At least five members of the frontiering Hardeman family built and operated mills from 1805 to the 1860s, from Tennessee to Missouri and Texas. Frederick Jackson Turner's "successive waves" theory of frontier advance was never more negligent than when it left out the role of the small-time mill owner and operator.

Shortly after the Civil War, grinders with metal parts for crushing the grain—rolling mills and steel burr discs—almost entirely replaced stone grinding wheels, except for large commercial wet milling operations. Both hand-cranked and belt-driven steel burr grinders became available at prices that many individual farm families could afford. At the other extreme, huge rolling mills operated by big corporate enterprises emerged. Thus the single-family–owned commercial mills powered by picturesque water wheels were quickly squeezed out of business, although some were able to hang on to meet their own family needs for a few years.

In modern times, several business people have started successful enterprises by advertising breads, pastries, and crackers of stone-ground flour made at some of the ancient mills. Pepperidge Farm, for example, operates the centuries-old Rose Mill at Millford, Connecticut. New Jersey, Virginia, and South Carolina all have operating mills that are two centuries old.

By contrast with the time-honored corn-grinding methods of pioneer America, the wet milling processes used during the early twentieth century stand out in sharp relief and illustrate the impact of technological change. The grain was steeped for twenty-four hours in a water and sulphur dioxide mix. The kernels were then cracked by the first milling and the germ was removed and crushed into oil meal. From this, salad oil and soap stock were separated. Then a naphtha process removed the remaining oil. Residue from the oil meal was sold to poultry raisers. The gluten, minus the germ, was further ground and flushed over long, sloping tables to settle out starches of various kinds. A centrifugal process removed more starch. (There were about a hundred kinds of starches.) Chemicals were combined with starches to get desired effects. For example, glues and postage stamp adhesive were derived from chemical and starch mixes, and hydrochloric acid was employed to change starch to glucose for use by such diverse establishments as hospitals and candy manufacturers. Glucose was turned into corn sugar by heating in vacuum pans, then pouring the substance ten inches thick on tables like huge trays of fudge. Blocks of this were put in chippers, and the "chip sugar" was used for making molasses, storage batteries, and alcoholic beverages, and for tanning boots. Glucose was also put into temperature-controlled converters, mixed with "seed" from a previous batch, and "split" to form powdered sugar and cattle feed. Gluten residues from various stages were fed to livestock or sold to distilleries. These few processes and products are little more than a sampling of what has replaced the small milling operations that once dotted America.

Until recent decades, rural American nostalgia was an old, creaking mill wheel, fed by a wooden trough of water from a nearby pond on a timber-lined stream. It was the "slosh, slosh, slosh" of overhead torrent hitting paddle buckets on the wheel rim, timed in syncopated rhythm with a "splash, splash" as buckets, turned under by the giant skeleton wheel's rotation, sent their cool, sparkling fluid pulsating into the tailrace below. There was a symphonic accompaniment of groans and creaks from the mill shaft straining against the crude, moss-covered housing and sturdy oak gears moaning under the load of transferred power, and a part crunch, part muffled roar from within the mill shed as hard, flinty corn kernels shattered in the grinding embrace between cap and nether stones.

The trips to the neighborhood mill were social occasions and often family affairs. Grownups prattled about crops and weather and local gossip, in chorus with the halloo and laughter of children fishing and swimming in mill-pond or stream or playing games or hunting squirrels among the trees. Kingfishers cackled noisily as they glided over the water, their bluish, white, and rusty bands reflected by the glassy surface. Shafts of sunlight, slanting here and there through a lofty canopy of verdant, leafy netting, caught the wildflowers and velvet moss and glittering mill-wheel spray and the turning, turning shadows in a kaleidoscope of contrasting hues. As surely as the mill stream waters warbled their liquid notes and passed over the wheel and through the tailrace, never to return, a new era and way of life blotted the old mill from its idyllic scene, banished it forever as an element in producing the country's staff of life. But art and song and a few careful historical restorations strive to keep the romantic wooden wheel turning in the stream of human consciousness.

CHAPTER XII

One Hundred Ways:
Food Uses

Bacon, corn bread, and coffee invariably appeared at every meal.

FREDERICK L. OLMSTED
on a trip to the South

Corn is the staff of life.

EDWARD WINSLOW,
New England, 1624

Most Americans now consider a receipt to be an acknowledgment of something having been delivered or received. The old-timers knew differently. A *receipt* was a set of instructions for preparing and cooking food, or in today's language, a *recipe*. Maize, or Indian food, was on nearly every pioneer table three times daily in one form or another. To the plate and the palate at all times of the year, it was as mother's milk to a newborn baby. Early nutritionists considered it a better winter food than wheat, owing to the greater warmth derived from its oil and starch.

Given its role in the production of eggs, milk, butter, and lard (the foundation ingredients of many recipes), as well as of meat, poultry, and cheese, corn would have been the best represented item in early American cookery if it had never so much as reached the dinner table in its natural state. But here too, whether as a vegetable in the form of roasting ears and hominy, whether ground for bread, cakes, mush, or pudding, or as a foundation element for beer and whiskey, it was unequaled by any other food in the pioneer diet. Corn is the country's national vegetable. Johnny cake, dodgers, pone, hoecake, ashcake, fritters, hominy, spoon bread, Indian pudding, hasty pudding—all of these spell c-o-r-n to us in no uncertain terms. True, this staple wore out its welcome at times, owing to monocrop monotony, but settlers, though tired of it, were tied to it and dared not keep it off the menu. There were no famines in pioneer America as there were in Europe, and the chief reason was corn, the most certain and most bountiful crop known to the western world.

Down through the centuries, trees fell, teams pulled, seeds dropped, and hoes chopped as the maize culture moved ever westward in answer to the hunger of both human and animal immigrants to the New World. Harvesters gleaned, mills whirled, stills boiled to satisfy pioneer appetites. Throughout American history there has been such great variety in the preparation and use of corn as food that no account of the subject can ever be complete, and a full treatment of what is known would require an entire volume.

The first fruits of the cornfield were not fruits at all but the soft culms, inner parts of young stalks. These were by-products of pulling suckers, the side growths from the bases of main stalks. The hearts of these were removed and eaten as asparagus is consumed today.

It was as if they could not bear to wait for the crop to ripen. Indians and settlers alike went into their fields and plucked tender ears no larger than an adult's finger. Usually they took these from stalks that had more than two ears so that the crop of mature corn would not be hurt. Some early agronomists believed that this helped the remaining ears to grow larger. The immature corn was eaten as a vegetable, cobs, kernels, and all, cooked or raw. As a rule, the pioneers boiled these ears and ate them with butter. Another practice was pickling them in brine or apple cider vinegar for winter use. Pickled baby corn is an expensive delicacy today.

Most crops—wheat, barley, rye, potatoes, fruits—were harvested at one time or as individual plants or fruits ripened, and that was that. Not so with Indian corn. There were suckers, baby ears, roasting ears, semihard grains, and mature ears, all with their own seasons. No stage was welcomed more at the pioneer table than the milky-grained roasting ear harvest.

Typical of a number of aspects of corn culture, settlers' food habits were blends of Indian innovations and pioneer improvements. So it was with roasting ears, or green corn, probably so called because they were gathered when the shucks were green. Ears were picked when grains were full but still juicy and milky as tested with the thumbnail. Farmers soon learned to tell the ripeness stage by feel without tearing the husk. Frequently, they ate green corn raw from the cob. Since about half the sugar turned to starch if the corn was not cooked within one and one-half days after picking, some gleaners rushed with their fresh plucked years from field to fireplace and ready-boiling pot.

The principal Indian method of cooking green corn, widely copied by white and black settlers, was to leave the ear in the inner husk and roast it in ashes or under a shallow layer of earth, with hot coals raked over the top. Another popular way was to strip off the shuck and lean the ear against a stone or an iron bar before the coals. It was turned and cooked to a light brown, coated with animal fat, marrow, or butter, salted, and eaten while held in both hands by skewers or by the stem and tip of the ear.

Boiling green ears was mainly a contribution of the Europeans, since few Indian tribes had the equipment to boil water. The cook of the cabin suspended a heavy iron pot of water on a lug pole or pothook over the coals. Cooking time in the steaming kettle was about twenty minutes, and the ears were eaten while hot in the same manner as when roasted. (The recommended cooking time is much less today.) Pioneers found this method of cooking particularly appealing since they first stripped the ear of its husks and removed worms and bugs before their juices permeated the surrounding grains. So simple was corn on the cob to prepare, whether roasted or boiled, that no receipt was needed, yet it was so basic as to have been probably the most popular food in early America during its season. "Certainly the greatest delicacy that ever came into contact with the palate of man," said a British observer in the 1820s. "I defy all the arts of French cookery . . . to produce anything so delightful." Nevertheless, the table manners of the American frontier, particularly in the eating of this delicacy, shocked European observers such as Frances Trollope.

By no means was all fresh corn eaten from the cob. Much of it the settlers cut off either before or after cooking to prepare it with butter or meat or fix it in a variety of other ways. Succotash, a mixture of corn and beans, they copied name and all from the Narraganset and other Algonquin Indian tribes of the Northeast in the early 1600s. It has been popular ever since. Other fresh vegetables, squash, onions, and peppers were added to the blend as desired. In all areas, fresh corn was creamed and fried like pancakes. Southerners ate it with syrup or honey and butter and called it fried corn. For a similar dish of fried corn and batter mix, New Englanders used the term *corn oyster*, because of the irregular, oysterlike shape of the cake. Fritter was another name used for the fried cakes.

Green corn pudding, or hasty pudding, was ever popular in early American times and was immortalized in nostalgic narrative verse from overseas by American diplomat, Joel Barlow, in 1793. A New England pudding recipe of ancient vintage is representative for the dish. "Take of green corn twelve ears, and grate it, to this add a quart of sweet milk, quarter of a pound of butter, 4 eggs well beaten, quarter of a pound of sugar, pepper and salt as much as is sufficient; stir all well together, and bake four hours in a buttered dish."

It was further flavored with various things: grasshoppers (used by many Indian tribes), fruits, and berries. Hasty pudding was considered "a rich and luscious feast." Many cooks, instead of grating, made a knife cut lengthwise through the middle of each row of grains, then cut from the cob with a knife.

In a field of corn all planted at the same time, the roasting ear or green corn period might extend to five or six weeks. Staggered planting times for different patches lengthened the season to as much as ten or twelve weeks. Still, it was desirable to have green corn for the bill of fare during the other nine months of the year. All through the colonial and early national periods, soldiers and braves trudged off to war and hunters and fur traders penetrated the backlands carrying bags of the lightest and most nutritious food to be had. An indispensable item in the duffel of practically every such wanderer was dehydrated, or parched, corn. English Army Captain Buttler told of the importance of dried maize to British and Iroquois soldiers in the 1710 invasion of French Canada. Peter H. Burnett, organizer and first wagonmaster of the Great Migration to Oregon in 1843, influenced countless overlanders by his widely publicized advice to take parched corn, raw corn, and cornmeal (in addition to corn in another form—forty pounds of bacon per person) on the long trek across the continent.

Parching was a very necessary preserving technique long practiced by most tribes and quickly picked up by the settlers. They cooked the corn and either dried it on the cob for winter use or cut it from the ear, dried it, and powdered it or left the grains whole. It would keep indefinitely and could be mixed with hot or cold water before mealtime. Some travelers mixed dried maple sugar with powdered corn to vary the taste along the trail. When parched grains were recooked they had much of the original flavor of fresh, green corn. The New York *Commercial Advertiser*, seeking to promote a means for winter sale of green corn, published a recipe for parching in the early 1830s.

> How to have Green Corn in Winter—Of all the productions of summer, there is hardly one more nutritious and palatable than green Corn. . . . It may be so prepared as to be as good and palatable in the winter. . . . The Corn must be taken from the stalk when it is full in the

milk, or in that state in which we generally use it—the husks stripped off, and the ears thrown into a kettle of boiling water, when it may remain half as long as you would boil it, for the present use. This will harden it so that it may be easily taken from the cob. It may then be spread in the sun till it is thoroughly dry; in preparing it for the table it may be soaked from twelve to fifteen hours, and boiled in the same water. But care must be taken not to boil it too much, as that will make it hard and diminish its sweetness.

Many Indians and settlers parched their green corn, either before or after husking, by simply drying it in the sun or near the fire, then hanging the ears for future use.

There were several other methods of preserving green corn. Women and children, who customarily did most of the cooking chores, boiled it, cut it from the cob, and salted it in crocks by pouring alternate layers of corn and salt, adding enough water to cover, and holding the grains beneath the liquid by placing plates on top. Another way was to pickle with vinegar. By the early to mid–nineteenth century, canning techniques were becoming widespread, and corn, as well as other vegetables and fruits, was preserved in nearly fresh form for the long winter season.

There was yet another phase of harvest in the fields of Indian corn before full maturity. Between the milky and dry kernel stages, settlers grated the semihard corn on curved tin gritters. The coarse grits which resulted were cooked in a variety of ways, but usually as a soft, sweet meal.

Grits was a term most often used to describe *hominy*, a word probably taken from one or both of the Algonquian expressions, *rockahominie* and *tackhummin*, the latter referring to grinding. Originally, Indian hominy may have been a recooked, parched corn. But for pioneers through the ages from Plymouth to the Plains, it referred to two means of processing and softening hard kernels for human consumption.

Mature, dry corn, especially of the flinty kind most used for making hominy, was much too hard for human teeth. Indian ingenuity had solved this problem in a manner promptly copied by the colonials and applied for centuries afterward in the making of hominy. The two types of hominy were whole, and hominy grits or pearl hominy. For the first, whole grains were soaked in lye (caustic potash, or potassium hydroxide) extracted from wood ashes. Early American fireplaces were often built with ash pits, and stoves were made with ash hoppers. J. A.

Graves, a California pioneer, described the lye-making process which, with slight variations, had prevailed for centuries. Farmers made hoppers or troughs of boards, tilting them downward, and providing drain holes at the lower ends. They dumped ashes into these outdoor hoppers and placed jars under the drain holes to catch the brown lye, or quicklime, as it was leached out by rain water. The lye was used to make soap, and it was the active ingredient in the essential process of soaking hard corn kernels to free their tough skins, or pericarps. Multiple washings removed the lye after many hours of soaking, and the loosened pericarps were separated from grains by rubbing with the hands. A final bath washed away the skins. In late pioneer times, cooks loosened corn skins by boiling the kernels in bicarbonate of soda and water.

Rural folk usually boiled the swollen, whole-grain hominy and ate it with salt, butter or animal fat, and meat; or they cooked it with butter and sweetened it with maple syrup, sugar, molasses, or honey. A traveler in the southern Appalachian foothill region of the early 1700s was served "Hominy toss't up with Rank Butter and Glyster sugar." Hominy is closely associated with the South, where, along with pork, it was a staple in the diet. But it was important in all areas of pioneer America, as seen from the presence of the hominy block, or samp mill, on nearly every farm. This simple but effective mechanism, as previously seen, was used to pound moistened, mature corn kernels into grits, or pearl hominy. The hulls were washed away and the finer meal winnowed out or sifted through a bolting cloth or a perforated deer hide. Women and children then prepared the coarse grits in a variety of ways, both salty and sweet.

Hominy grits, usually of white corn, have been called "the potatoes of the South," so heavily have they been relied upon for starch in that section. They are "the sole food of the negroes," wrote an early traveler. Such vendors as Dutch Molly and Clio sold boiled hominy in the streets of cities along the American seaboard. Grits were cooked as soup, porridge, or gruel, fried as cakes, or spooned on to the plate as vegetables.

Rural chefs simmered hot hominy over slow heat for many hours, enriching it with butter and perhaps cream, and salting and sugaring to suit the taste before serving. Southerners were notoriously fond of hominy and meat gravy. Other popular dishes were hominy fritters— grits mixed with milk, flour, and a beaten egg, and fried—hominy muffins, and hominy mush. Grits were eaten as a main part of the

meal or sweetened and served as dessert. An old-time breakfast combination of eggs, ham, and hominy grits is still popular in the South.
Slaves and poor whites of this section of the country were much more
inclined to cook hominy and other corn preparations with a variety of
vegetables and meat all together in one pot, much like chop suey in
Goldrush California. One such dish was Hoppin' John, a slave con-
coction of grits and peas. There was a simple reason for one-pot meals:
poor folk often had only one cooking pot.

Because cracked corn, with the oily germ included, had a tendency
to develop a strong, rancid taste after several days, it had to be processed frequently. The custom was to "limber up" the hominy block
daily, cracking only as much corn as was needed for the day and the
following morning's breakfast.

A sizable percentage of the mature, dry-ripe corn crop was shelled
and ground into meal, flour, and grits. Although more people preferred
white ground corn for the table, they often used yellow and the less
common hues. White meal was generally favored in Massachusetts
and Mississippi, for example, and yellow in Illinois and parts of Indiana. The best tortillas in the Southwest were made of blue corn, despite the moldy look. Ground corn was prepared in so many ways as to
defy full description, nearly every family having its own variations in
recipes. In practically all of these, it was mixed with water, milk, buttermilk, or clabber. The mixture was rather dry for breads and cakes,
medium moist for mush and puddings, and thin for porridge, gruel,
chowder, and soup.

If bread was the staff of life, it was a corn-bread staff in America
until "wheaten" bread began to catch up in the middle of the nineteenth century. And in the South, corn bread kept its lead until long
past the Civil War.

The simplest corn bread recipes called for moistening meal, adding
salt and perhaps sugar and a flavoring, then baking until brown on
top. Yellow bread, it was commonly labeled, whether made from yellow or white kernels. A host of ingredients could be added: bacon
grease, butter, berries, bacon bits, and for "cracklin' bread," cracklings
left over from rendering lard. Slaves usually cooked their corn bread
without adding any shortening.

Pioneers always preferred their corn bread, like hominy, fresh and
hot. They cast it in many molds—rectangular pans, round tins, muffin
rings, griddles and sheets, hoes, and cast iron corn gem or corn stick
pans with depressions either smooth or shaped like ears of corn halved

lengthwise. Name changes often accompanied changes in the shape, mold, and cooking methods. All recipes were "wild" in the sense that there were endless variations to suit tastes and the local availability of ingredients.

Johnny cake, probably of New England origin, spread to much of the South and Midwest. Its name may have been a modification of *journey cake*, for those who made it to eat on the trail. There was scarcely a recipe for it at all: cornmeal and water or milk, perhaps a little salt or sugar and fat, mixed and baked flat before an open fire. (If salt was missing, hickory ash would have to do.)

A similar corn bread was pone, probably from the Algonquian term *apan*, meaning baked. This southern and midwestern staple, heavily depended upon by slaves, was sometimes baked, sometimes boiled like a dumpling. Ash-cake was likewise a small loaf of corn bread wrapped in corn husks or cabbage leaves and baked in hot ashes. Some pioneers, lacking other pans, removed the handles from their field hoes and molded small corn bread hoecakes on the blades. These were then placed close to the open fire until the cakes were cooked. The hoe blades could be subsequently quenched in cold water to restore temper for field use. Henry Thoreau first made his hoecakes of corn and salt but eventually "found a mixture of rye and Indian meal most agreeable."

Another small and very popular corn bread was the dodger. Abe Lincoln recalled from his early years that his family ate wheat cakes on Sunday and dodgers the rest of the week. The corn dodger may have acquired its name because it was so hard that it was dangerous to be hit by one. A typical proportion for all these small corn bread types was four parts meal to three of liquid by volume.

Hush puppies were patties of corn and hash or perhaps corn batter spooned into deep fat and fried. Some southern hush puppies were fried in fat left over from cooking fish. Cooks often used onions and peppers in these and other corn cakes for flavoring. The name *hush puppies* is of uncertain origin. It may have been assigned because they were used to quiet whining dogs, or possibly *hash* puppy became *hush* puppy.

The list of other corn bread and cake variations is a long one. Flat cake, water cake, and sad bread were similar in consistency to the johnny cake. Berry bread was corn bread with huckleberries, raspberries, strawberries, or blackberries. And there were nut cake, pumpkin bread, salt rising bread, pound cake, corn sticks, and corn muffins.

In time corn bread recipes included eggs and oil. A typical recipe called for the baker to mix together one pint milk or buttermilk, two beaten eggs, and one pint cornmeal. Add a teaspoon of soda, a little salt and sugar to suit taste, and a flavoring such as one-quarter cup of melted bacon fat. Bake in a greased, heavy iron skillet until the crust is light brown. Since it required no yeast, the corn dough was much easier to prepare than wheat bread. Baking powder was a latecomer to the American pioneer world, but some settlers made a substitute from corncob ashes and sour milk. In any case, since corn has little gluten, unless it was mixed with other flours, leavening would have no effect.

Rye, rice (in the South), and wheat flour were eventually incorporated in corn breads. A mix of corn and wheat flour came earlier in the North (by 1800) than in the South. Wheat flour became much more tightly compacted than cornmeal, and a leavening became necessary to prevent every loaf from turning into hard tack instead of bread.

Cold leftover corn bread was rarely wasted. A typical settler would break it into a bowl and eat it with milk, buttermilk, or hot gravy, fry it like hash, or crumble it to make corn-bread pudding. Broken corn bread was also a standard item for poultry stuffing. If it became too stale, sour, or rancid, the leftover bread was fed to hogs.

Cornmeal pudding was perhaps less a favorite than hasty pudding, but like the other cornmeal dishes, it was available twelve months of the year. Spoon bread and batter bread were other names for it. The recipes were just as open and flexible as for corn bread, whether meat or sweet, and the proportions of ingredients similar except for increased amounts of milk. It was customary to serve pudding at the evening meal, and there were many times when it made up the entire meal. A favorite way of eating it was with molasses or honey.

Porridge, similar to the Indians' *suppawn*, was a dish much like pudding but thinner in consistency. It was made by boiling broth or water with milk and thickening with finely ground cornmeal. Said an early nineteenth century writer from New York, "Porridge . . . is quite sufficient for two of the meals" in a day, so long as meat is available for the third. Corn porridge from a clam shell spoon was familiar to countless settlers. One advantage of corn pudding, porridge, and mush was that they were good sources of starch that could be prepared in less time than bread. The breakfast counterpart (both breakfast and supper in the South) of pudding and porridge was cornmeal mush, called cush cush in some areas. It was eaten hot with sugar, molasses, honey, or maple syrup when such sweeteners were available, and with milk.

Sometimes it was flavored with butter, bear grease, nut oil, or crack-lings. If there was leftover mush after all stomachs were filled, it was sliced when cold, and fried in bacon grease for another meal. Mush was a favorite throughout pioneer America but particularly in the South and among the Dutch settlers of New York, who for long periods ate little else.

Every farm needed an accessible corn grinder or cracker, since ground corn, like hominy, quickly developed a rank taste. Keeping it in a cool place helped, but like Ben Franklin's fish and guests, ground corn tended to spoil after three days.

The Southwest has been, since pre-Columbian times, the land of tortillas, tamales, posole, and atole. All were made from corn freshly ground on a metate. The thinness of tortillas, observed Josiah Gregg, was the test of skill. For tortillas, the Indian, Mestizo, and Spanish women patted the corn dough (moistening their hands at frequent in-tervals), placed the pieces on hot stone slabs or iron or copper griddles called comals, and turned the wafer-thin dough to cook on both sides. The tortillas were rolled into cylinders, folded into sandwiches, or overspread with preparations of meat and vegetables. Thomas Jeffer-son Green of Texas commented in 1843, that these corn sheets served as "knife, fork, and spoon" in conveying dripping delectables to hun-gry mouths. He might have added dish and napkin to the list. Tortillas and the mixes that they held were almost the entire diet of some peo-ples of the Southwest. As sailor Richard Henry Dana noted in the 1830s, they were a part of menus in far-off California. Anglo-America has adopted the tortilla in its native soft-dough form as well as in mod-ified chip-and-dip fashion.

The tamale was a wrapping of cornmeal dough around meat, vege-tables, or a combination of the two. The little loaf was encased in corn husks and boiled. A corn or hominy soup called posole was widely con-sumed, and atole, a thin mush of Indian meal, was also very important in the diet of New Mexico residents. They drank the souplike mixture with frijoles (beans) and chiles (red peppers). They "breakfast, dine, and sup" on this liquid corn mix, consuming more of it than the Amer-icans drink coffee, observed Josiah Gregg in the early 1840s. "El Café de los Mexicanos," Gregg called this dish of ancient origin. Still an-other southwestern creation was the paper-thin piki bread made of blue cornmeal and ashes.

Certainly one of the oldest corn preparations, copied by immigrants from the Indians, was to heat the grains of either yellow or white pop-

corn until they expanded explosively from dehydration. Popcorn became a universal favorite around the evening fireside all over America.
After 1840, when more and more fireplaces were boarded up in favor
of iron heating and cook stoves, the staccato cracking sounds shifted
to those metal focal points of social interaction. There were Franklin
stoves, sheet metal airtights, box stoves, potbellies, base burners, and
ranges. And the family circle reenacted the fireside scenes night after
night in home after home across the big-shouldered countryside: fea-
turing folk tales and accounts of family exploits, stories of war and
hunting, balancing of youngsters on knees of the elder members of
the circle, tobacco spitting, passing the jug of corn juice or cider, and
"pop-corn going off in the skillet, like the volleys of musketry we were
so soon to hear at Shiloh."

And so the list goes on—and on. Southerners rolled fish, meat, and
chicken in fine cornmeal before frying, and their migrations spread
this culinary practice through all the trans-Appalachian corn country.
Those succulent, green stalks of maize were a source of sugar when
chewed. Children in a sweet-scarce environment, where molasses,
honey, and maple syrup were never as plentiful as desired, found the
stalks particularly delightful. Some settlers made a molasseslike syrup
from corn stalks as early as the American Revolution. Later a technique was developed for making molasses from corncobs.

The people of New Mexico had a sugar problem of a slightly different type—the usual sweet-tooth condition, frustrated by very high
import costs, and unlike much of the eastern wooded area, a lack of
local honey or maple sugar sources. George R. Gibson, a soldier in the
Kearny-Doniphan expeditionary force during 1846–1847, told what
they did about it.

> The people of the United States are greatly behind the Mexicans in
> making cornstalk sugar, as it is a very common thing both here and far-
> ther south. The Santa Féans are now employed at it, the process being
> very simple. They take the young stalks, beat and mash them up with
> malls in large troughs, then press them in a press such as we generally
> have at cider mills, and boil the juice in kettles. Today I saw the process
> in all its different stages in three or four places in the upper part of town.
> The sugar is pleasant and put up in cakes with corn shucks, and for a
> long time was mistaken by us for maple sugar.

Malls were not the only means of crushing cornstalks for boiling
"miel," or sugar. According to Lieutenant William Emory, men, wom-
en, and children derived "the highest glee" from riding seesaws that

crushed cornstalks in hollowed cottonwood trunks. This process, a legacy from the Spanish and Mexican periods, continued into the 1870s. Meanwhile, the Union Sugar Company of far-away New York began making sugar of cornstarch from the kernels in 1865, and the process and product have been expanding commercially ever since.

Cornstarch processing was almost as old as the introduction of corn to the first seaboard settlers, whose ancestors had made starch from other grains since Roman times. The chief early American food use of cornstarch was as a thickener in puddings, sauces, and gravy. Later it became an ingredient in soups, baking powders, and batters, although its primary value has been in the nonfood industries.

It was long known from observations of cooking residues that corn had a considerable oil content. Individual attempts to make practical use of corn oil during the pioneer period met with little success, but by the Civil War era, processors were extracting vegetable oil from the corn kernel germ on a commercial basis. One hundred bushels of grain produced about sixteen gallons of oil, but the separation was expensive. That corn oil was a good substitute for animal fat, much better than oils from the other common grains, was understood, and since the 1860s its uses for cooking and salad oil as well as for industrial purposes have been constantly on the increase.

Perhaps a fitting conclusion to corn's array of human food uses is a Missouri recipe for—of all unlikely spreads—corncob jelly. Break a dozen dried red cobs, cover them with hot water (a plate will keep them submerged), and let stand for ten to twelve hours. Boil for one-half hour and strain off the juice. Add a box of pectin to three cups of the juice, bring to a boil, and pour in three cups of sugar. Boil about five minutes or until it drips slowly from spoon, pour into jars, and seal.

It would have been too much to expect that corn would serve so wide a range of human dietary needs without some tarnishing of its reputation. There were those during the colonial period who believed it to be a cause of worms in the digestive tract. Irish famine victims spurned it in the late 1840s for fear it would cause their babies to be born yellow.

Pellagra has been by far the most serious corn-related diet problem in the United States. It persisted throughout much of the South in epidemic proportions into the twentieth century, although it has not been a serious threat since the 1930s. Pellagra afflicts people who are deficient in niacin (nicotinic acid, or vitamin B), and who are exposed to warm sunlight. Epidemics have occurred in the Balkans, Spain, India, and Egypt and other parts of Africa. The symptoms of the disease are numerous, often involving sluggishness, loss of appetite and weight, diarrhea, ulcerated tongue, redness of the mouth, mental depression, and fever, and nearly always including skin eruptions on neck, arms, legs, knees, and backs of hands. Fever, delirium, and prostration may be present in severe cases.

Because pellagra was clearly associated with a very limited diet of corn, salt pork, and molasses in parts of the American South, it has been widely believed that these food elements were direct causes of pellagra. Modern-day laboratory isolation of a number of vitamins and the successful treatment of pellagra with niacin and controlled diet clearly establish that none of the elements in the diets of pellagra victims, but rather the lack of vitamin B, actually caused the disease, just as lack of vitamin C caused scurvy, not the specific items in the diets of scurvy victims.

Without question, the ease of producing corn and the expense of other food items contributed to an unbalanced bill of fare, but the fault was not with corn. The very heavy corn-and-pork diets characteristic of the South caused susceptibility to numerous diseases.

Although settlers of the Old Northwest also relied very heavily on a hog-and-hominy diet, they also ate more wild game, berries, small grains, and other varied foods than did most southerners, particularly slaves and others of the so-called lower classes (free blacks, small farmers, and poor whites). Each person and each hog in the South, according to author John D. Clark, ate, on the average, about thirteen bushels of corn per year, or nearly two pounds a day. Adding to this the fact that each slave ate some 150 to 200 pounds of bacon and other salt pork per year, one can appreciate the importance of corn products as food, despite the reputed appetites of blacks for opossum and watermelon. Slaves were particularly vulnerable to cost-cutting in the menu. As an observer of the 1820s noted, "All the BLACKS on the continent of America, and in the West Indies, are fed upon corn in some one or another of the modifications." The fact that there were many modifications did not change the basic nutrients.

Among the modifications of corn as food, animal products were by no means insignificant. So diverse were the direct food uses of the versatile plant, so heavily dependent on it were the generations of American pioneers, that it would be easy to overlook the great grain's dominant role in producing meat, poultry, eggs, and the wide range of dairy products. All of these secondary forms of corn consumption were significant but none more so than the ever present bacon, or sow belly, as it was commonly called.

Americans are conditioned by their photo-media to visualize early settlers' livestock as blooded strains, pure-bred species hauling white-sheeted wagons, munching grain and stover, or kicking up billows of prairie dust with their thundering hoofs. Actually, with rare exceptions such as churro sheep and Arabian horses in parts of the Southwest, those domesticated beasts and birds that thrived on the corn crop were all mongrels for some 250 years after the founding of Jamestown. So thoroughly were they blended that scarcely a pure-blooded specimen could have been found in the entire land from the Atlantic to the Rockies and beyond. There never was a stampeding herd of white-faced cattle on the way to the early Kansas railhead, nor a team of blooded, white-stockinged Clydesdale horses pulling a Conestoga wagon across the plains to Oregon.

The Indians had no animals to work their cornfields, and as a matter of spiritual conviction they disapproved of feeding the sacred crop to animals. A convenient belief, perhaps, since maize was critical for human survival. When the tribespeople gave corn to white visitors for their mounts, as they did to William Bartram in the South, they did so as a gesture of great respect. New Mexico residents rarely fed corn to livestock. For settlers farther east, however, the feeding of crops to work animals was a good bargain, since the beasts provided a margin of surplus production over and above what they ate. Then too, much of the feeding of domestic animals provided protein enrichment for the settlers' diet, an enrichment which the Indians got from wild game. The more virgin land was converted to grain fields, the fewer the wild birds and animals, and the greater the settlers' dependence on domestic stock for food. Thus more corn created a need for still more corn.

Horses, mules, and oxen produced corn; corn produced horses, mules, and oxen. The feeding of these work animals became essential to the endless food chain. If a farm was successful, this process would generate a grain surplus, which would meet many needs beyond the simple refueling of the work force.

Domestic livestock, like humans, ate corn in many forms and at all stages of growth. Farmers took early advantage of the growing corn crop by feeding their animals the small green stalks pulled in thinning and suckering. Later they would cut matured green stalks for feeding fresh or for fermenting into silage. From November till the tender grasses of May, dried fodder or stover, which was as nutritious as hay, was very heavily relied upon to winter the stock. In much of the country, and particularly in the South, fodder or stover was far more important than hay as animal feed. Nearly half of the corn plant's food value was in its fibrous leaves, stalk, husks, and cobs. Although these were not very appetizing in dried form, they could be successfully fed. The settlers had a trick or two up their sleeves, such as crushing cobs and kernels together (the grain serving as an appetizer), and feeding fodder in cold and snowy times when there was no succulent, green food and animals were lean and hungry.

Even so, the beasts ate only the leaves and shucks of the stover, shunning the coarse but nutritious stalks. This wasted half the food value of the fodder. Advice from farm almanacs came to the rescue. By chopping the stalks in one-quarter inch to one inch lengths (sometimes with a hatchet in a wooden trough, as Thomas Jefferson advised), and soaking them in hot or salty water, the settlers could transform them from coarse refuse to edible feed. Livestock often preferred this soaked fodder to dry hay. As a last resort for finicky animals, the farmers mixed cornmeal or even a little molasses, when available, with finely chopped fodder before feeding. It worked like honey on stale bread for children.

Farm dwellers fed whole stover to their animals by scattering it on the ground in the barn lots, dumping it in large feed bins (particularly in the stalls of stables for horses and mules), or turning the animals loose in the areas where the fodder stacks were located. Green nubbins, or stunted ears, they fed by tossing on the ground in feedlots, but chopped fodder and whole or ground grain posed a different problem. Except for poultry feeding, the scattering of kernels upon the ground would have caused much waste. Thus, traditionally on American farms, pigs ate from simply made V-troughs, and horses, mules, cattle, and sheep from troughs, boxlike bins, or round tubs made of half barrels. Then too, there were self-feeders of various types where meal, whole grain, or even ears of corn dropped down automatically from vertical hoppers and replenished what the animals ate from the troughs at the bases of the hoppers. This principle was used for both livestock and poultry. Another type of feeder that minimized waste was a canvas bag strapped over the muzzle of a horse or mule. This feed bin moved with the animal until it was emptied.

There was no question about the animals' taste preferences. They liked grain better than fodder, and they preferred corn to any other grain. Farmers fed corn kernels, green or ripe, shelled or in the ear, whole grain or ground meal, wet or dry, and at times recycled. Of all the stock and poultry feeds, corn grains were best for fattening and providing work energy.

Horses and mules, and occasionally hinnies, were the only species of farm animals fed exclusively for work. Unlike other stock and poultry, these beasts were not acceptable to the settlers as human food, and they produced no eggs or dairy products. They had to carry their weight in the fields.

Field and road work called for energy. Fodder, hay, and grass were useful to tide animals over slack periods, but grain was a necessity for working animals. While oats, rye, barley, and wheat bran were considered highly useful, agricultural writers in early America consistently maintained that horses fed on a diet made up largely of corn were superior to others in speed and endurance during hot weather or cold. Race horses in the early nineteenth century were called cornfed dandies. From the 1790s onward, many southern farmers came to prefer mules over horses. The long-eared creatures could withstand heat better and were less likely to become foundered from overeating.

Mules and horses were well equipped with sturdy teeth. They could cope with the driest of fodder and the hardest of kernels. Before the development of single-family corn shellers in the 1840s, most farmers simply dumped ears of dry, mature corn into the feed bins of these animals. The crunch of flinty kernels between rows of equine molars echoed throughout the stable barns. Later it became more common for farmers to shell and even grind the grains for horses and mules. Army rations for mules and horses were usually shelled corn to avoid transporting cobs. A working horse or mule needed, on the average, sixteen to twenty pounds of grain and eight to ten pounds of hay or fodder daily.

American farmers did not plow by horse and mule alone. They relied on working cattle, known as oxen, in all times and areas. But they fed cattle for much more than work. These animals were efficient biological machines for converting corn and other feeds into meat and milk. The latter was consumed directly or processed into butter, cheese, and cottage cheese, better known as smearcase (smear cheese).

Through the long months from late fall to early spring, cattle ate large amounts of fodder. Milk production was improved by feeding chopped fodder soaked with water. Still better for milk cows was silage, which provided the advantages of a green feed during the cold season. Farmers were advised to feed corn kernels or ground corn for better grades of meat and dairy products (especially for yellower butter) and for sustaining oxen during times of heavy work.

Cattle could not chew and digest whole-grain corn as well as could horses and mules. This was not a problem when the grain was milky, and green ears were pulled from stalks for the bovines to eat—husk, grain, cob, and all. During the seasons from October around the calendar to July and August, however, green corn was not available. The pioneers achieved much better results by feeding ground corn, or Indian meal, to beef and dairy animals. But grinding was expensive, and by observation, the settlers soon learned that there was another way— recycling in the barnyard.

Eventually this became known, with some variations, as the corn-cow-hog-poultry cycle. The kernels which were unchewed and undigested passed through the digestive tracts of the cattle into excrement. Hogs then scavenged this waste corn from the manure and became fat on it. In turn, those particles of grains which passed undigested through the hogs were picked from their manure by the sharp-eyed and sharp-beaked chickens. European visitors, often critical of American ways, marveled at this combining of minimum labor and reduced waste with maximum production. The cycle has been widely used ever since. Some weeks before butchering time, stock feeders took their hogs off the scavenger diet and fed them "unused" corn.

Hogs, like other domesticated meat animals in early America, spent much of their growing time (as contrasted with fattening time) roving and foraging the range. As will be seen later, they created fencing problems and feuds between neighbors when their ever-gnawing hunger led them into grain fields and gardens. So voracious were the appetites of these porkers that their names have become synonymous with overeating and with greed in general. What the rooting beasts ate in their gorging, fattening stage was mostly corn. They ate "slop," yes, and all the other edibles that omnivores crave, but more than anything else, they ate corn in all of that bountiful grain's forms and stages. Sometimes, in fact, these waddling, wallowing pigs were more ungainly than usual as a result of food-caused drunken spells. For "best fattening," a farmers' almanac of 1821 advised mixing ground corn and water and allowing it to ferment for two to three weeks before feeding it to hogs.

A large percentage of pork producers cut costs of both harvesting and feeding by letting their hogs do the harvesting directly from the cornfields. This practice, as previously discussed, was known as hogging down the corn.

Whatever the process, and there were many variations, hogs efficiently converted feeds into some of the most favored products of pioneer America: pork, ham, bacon, lard, cracklings, souse or head cheese, and pigskin. Short of Texas and the Great Plains, pork was more important than the meat of any other animal in rural areas of the country.

Sheep and goats, although less significant than other livestock, were also consumers of corn. Grain-feeding of sheep grown primarily for wool was not common, but corn was the best fattener of animals raised for mutton. Unlike cows, sheep ate whole grain corn and digested it thoroughly.

Poultry, too, were well equipped to digest whole kernels of the golden grain. Uniquely equipped, in fact. Stuffing their craws with small stones and commercially crushed oyster shells as well as corn, they created natural stone corn grinders. Very young fowl were fed cracked grain. Chickens, ducks, geese, and turkeys preferred corn to any other grain. And pioneers preferred corn-fed birds for both their rich yellow fat and deep orange egg yolks.

Women and children usually did the feeding of chickens, scattering grain broadcast in the poultry yard, then standing by to shoo away sparrows and other such uninvited dinner guests. It was not necessary to wait long, as the fowls stuffed their craws in a few moments, then let their gizzards take over the more leisurely grain grinding before digestion. It was a simple, cheap feeding process—no grinding, no mess, no costly feed containers, no waste.

Avoiding waste was often crucial in marginal frontier economies. Few better examples of shunning waste can be found than the corn-cow-hog-poultry cycle. Perhaps this economy was equaled in the double-use of grain mash, or "distiller's grain," left over from whiskey distilling. Cattle, and particularly hogs, ate huge quantities of corn mash salvaged from the thousands of "stills," or distilleries, that dotted America. The resulting savings amounted to untold millions of dollars over the generations. How symbolic that pork, the most important meat of the pioneer American corn belts, was so closely related to that premier American drink, corn whiskey.

CHAPTER XIII

One Hundred Proof: Pioneer Spirits

Corn has, "like some other of the best gifts of the Deity, been perverted to base and injurious uses."

EDWARD ENFIELD, 1866

My grandaddy made whiskey.
His grandaddy did too.
We ain't paid no whiskey tax
Since 1792.

We just lay there by the juniper
While the moon is bright;
Watch them jugs a'fillin'
In the pale moonlight.

From an old folk song

Indian summer had settled over western Pennsylvania. It was early fall of 1794, and along the rolling Alleghenies foliage was ablaze with color, licking like tongues of friendly flame against clear blue skies. But not all was at peace in this harvest season of the farming frontier. Incredibly, United States Secretary of the Treasury Alexander Hamilton himself, in company with Commanding General Henry "Light-Horse Harry" Lee, was invading with nearly fifteen thousand armed militiamen. Their object? To put down a backwoods rebellion and serve notice that distillers must pay the cash excise tax of seven cents per gallon that the federal government had levied on producers of whiskey.

Several hundred miles from the rum-producing coastal towns and too far away to ship their cheap grain economically over the ranges of mountains to tidewater centers, the frontiering settlers had taken to converting corn and rye to whiskey. This high-value, low-bulk commodity went far to solve their problems of trade with distant points. Albert Gallatin, country-store owner and political representative from western Pennsylvania, said his people had no choice but to concentrate "the greatest value in the smallest size and weight." One pack animal plodding across the highlands could carry up to six times the value in whiskey that it could take in corn. The beverage was becoming popular on the coast, thanks to a cutback of the molasses and rum trade in the revolutionary era and afterward, and the whiskey-filled keg, demijohn, and "Black Betty" jug moved freely in frontier trade.

Hamilton's "soak the poor" tax was not the first levy that these backwoods bootleggers had been ordered to pay—nor would it be the last. The state had earlier imposed such a tax, and before the Revolution, most of the colonial governments had also. However, it was the largest by far, and the first demanded in cash.

The independent "Spirit of '76" was still fresh, and it flared up spontaneously over taxes on the spirits of the 1790s. Stills, after all, were as valuable as acres of waving corn and rye. Outraged backwoods dwellers called mass meetings, and committees of "Whiskey Boys" were organized. Scotch, Irish, and Scotch-Irish settlers, some of whom had left the old country and brought their whiskey-making talents to America partly because of such taxation by the English government, vented their wrath on revenue agents as well as fellow producers who had paid the hated duty. But dealing with fifteen thousand troops was a different matter. The Whiskey Boys' open rebellion collapsed in frontier Pennsylvania as did forcible resistance in Kentucky, and uprisings failed to develop elsewhere in the West, although not because moonshiners were absent in other areas. William Byrd had observed lively whiskey distilling activity in the Virginia–North Carolina frontier region by 1728. Lacking accessible mills, the Scotch-Irish who had migrated down the Great Valley of the Appalachians to the southern backlands made small querns to grind their sprouting corn for "licker."

Actually, there had been very little effort to enforce the whiskey tax in the South and Southwest probably because, unlike Pennsylvania potions, "mountain dew" from more remote areas offered less competition to rum producers of the northern seaboard states. But President Washington, a whiskey distiller himself, and Hamilton sought to make an object lesson for all westerners by firmly enforcing the federal law in Pennsylvania.

The end of tarring and feathering of United States revenue officers by no means brought a halt to moonshining, however. Faced with a burdensome tax, a number of the Pennsylvania whiskey producers switched to hog and cattle raising. Others picked up their businesses, mill, still, and skill, and flatboated down the Ohio River to the new state of Kentucky, where they reopened shop in a county named Bourbon (for the Bourbon monarch of France, who had helped America in the Revolution). Hence the famous bourbon whiskey. There were well-known distillers ahead of them, such as the Reverend Elijah Craig of Scott County, Kentucky, who had been producing good quality whiskey since 1789.

At the urging of President Thomas Jefferson, who, incidentally, began making corn whiskey at Monticello in 1813, the oppressive tax of 1791 was repealed eleven years after its passage, simultaneously crippling the American rum industry and boosting whiskey production. There would be other whiskey excises, state and federal, spanning most of the decades to the present time. Revenue inspection and collection remained a way of life, as did the art of concealing stills and running blockades. Until the advent of mid–twentieth-century surveillance and detection technology, it was almost a game—a serious game, to be sure, but a contest nevertheless. Sometimes it involved unofficial sporting rules as in the case of hounds and hunter and foxes. There were thousands of humorous incidents and not a few tragic ones to spice the lore of the moonshiner. Bootlegging motives went deeper than the big profits and ready cash that came with success. Excitement and a feeling of triumph in outwitting the "revenooers" and personal pride in the product must have been important elements as well.

Frontier store owners gave more goods for cash on the barrelhead than for credit purchases. Corn whiskey was cash *under* the barrelhead. Furthermore, it was a universal home remedy and the national drink in pioneer times. It came to America as a full-blown distilling technology with the rye-whiskey producers who arrived as early as the 1600s from "proofing" grounds of England, Holland, Wales, Scotland, Ulster, and Ireland. Although there may have been earlier colonial distillers, Dutch Director William Kieft of New Amsterdam was the first known whiskey producer on the Atlantic Seaboard. He made grain spirits on Staten Island in 1640 and also levied the first tax on such beverages. A knowledgeable European traveler wrote of the "fine spirits" distilled from corn in back country New York and Pennsylvania as early as the 1740s. The name *whiskey* probably comes from the Gaelic-Irish word "uisgebeatha." Scotch-Irish immigrants, who brought the name across the Atlantic, included some of the most famous distillers in America.

In the vast western lands, and particularly on the trans-Appalachian frontier where the Scotch-Irish settled, whiskey quickly found its home, its sanctuary. There was freedom (for a time) from excise tax officials, a growing market of thirsty patrons, and corn, a superb brewing grain that could thrive and outproduce any other over a vast geographical and climatic range.

Not that potent corn drinks had failed to develop before importation of whiskey know-how and the simultaneous opening of vast new corn lands. Distilling was unknown to the Indians, but fermentation was not. Pulque from agave plants, juices from a variety of nuts, fruits, and trees such as maple, and deadly extracts from wild tobacco and jimsonweed had induced sensations from mild inebriation to wild hallucination, even death.

Among the juices fermented and drunk by the Indians, corn was most important. They prepared it in several ways—soaking cornmeal or green stems in water; mashing green, milky grains to a thick liquid; producing enzymes by moistening and sprouting kernels; or chewing soft grains and adding this fermentation starter to corn liquid or mash, much as mother of vinegar was used in apple cider by the colonists. Time did the rest. While the alcoholic content was rarely over ten percent, it was enough to produce violent intoxication if drunk in large quantities. And there were such times, notwithstanding the jest (by journalist-historian Bernard de Voto) that the Indian civilization deserved to perish because, when their corn spoiled or fermented they threw it away.

The colonists came along with a mechanical advantage, having known both fermenting and distilling techniques in Europe, and they took to corn like a drunk to a free jug. They preferred to make near beer, one of the common fermented drinks, from blue grained maize, allowing it to sprout and grow until the shoots turned green, then washing it daily to keep it from molding, and consuming it after a week or two of ripening. Blue corn was considered as good as barley for malting. The settlers also made near beer from broken corn bread.

There are many records of corn ale and beer from early times. Thomas Harriot wrote in the 1580s that malt from corn was brewed into ale and combined with hops to make good beer in Virginia. Later, hops would be one of the first crops planted on frontier farms. Swedish settler Peter Lindstrom of the Delaware River colony wrote in 1654 that black, blue, brown, and pied corn were brewed into beer. In 1790, a New England writer provided directions for making a "good malt." Bury corn two to three inches deep in earth, and cover with a loose mold. In ten to twelve days it will sprout. Take it up, wash and fan away the dirt, and put the sprouted kernels immediately into a kiln.

Hector St. Jean de Crevecoeur noted that some beers were brewed with roasted corn. Others were made of part cornmeal as shown by a beer recipe from a farmers' almanac of the early 1830s.

> TO THE LOVERS OF GOOD BEER
> Put two quarts molasses into a keg with ten gallons of cool water. Boil two ounces of alspice, two ounces of ginger, two ounces of hops, and half a pint of *Indian meal* in two or three quarts of water about an hour—strain it into the keg while hot, add one pint of yeast—shake it well together—stop the keg nearly air tight, and let it stand for twenty-four hours, when it will be fit for use. The whole expense of this quantity will not exceed three shillings.

Most large plantations had malt sheds and breweries for making beer. Sweet juices extracted from cornstalks were consumed before fermentation, as in California during the early 1700s, or used to brew a kind of brandy or rum. Even corncob wine was not unknown. The settlers made many other drinks from their various fruits, domestic and wild, particularly wine from grapes and berries, hard apple cider, and apple jack distilled from cider. Apple production was slower to build up than corn since the trees took years to mature, but in time cider replaced much beer making.

However, the genuinely hard liquors, rum and brandy and particularly whiskey, were the drinks that hit the New World with a culture-crumbling impact. If rum was the spark that ignited the slave trade with Black Africa, so corn whiskey was the lash that drove the fur trade with Red America. The Indians' own heaven-sent maize had boiled up in their faces. Those first masters of the marvelous crop became its slaves when its most lethal ingredient was released by Old World techniques and turned against them by Old World traders.

Patterns of trade with the Indians were remarkably similar, no matter the place or time. Like a mechanical gadget that ran more smoothly when oiled, the fur trade machinery was lubricated with liquor. This potent drink was no respecter of races in its deadly effect, but for reasons over which scholars still argue, the Indians seemed to have especially tragic reactions to hard liquor. Distilled alcoholic drinks rank, along with diseases, horses, axes, weapons, and traps, among the white contributions that had the greatest impact on Indians. Whiskey, dispensed freely before trade negotiations began, paid huge dividends in lowered prices that white traders had to give for furs. This became standard practice when the colonial period was yet young.

Both British and American governments tried, although somewhat feebly, to protect the Indians from greedy, ruthless traders. Frontier peace and economy were as much the motives as was Indian welfare. Such attempts failed through the centuries largely because settlers could vote, influencing policy and officials, while the Indians could not. In 1795 the government established a system of trading houses in the West. It was a kind of post-exchange setup, a federal monopoly designed so the native peoples could deal with the government and avoid private traders. After financial failure and strong opposition from trading outfits such as John Jacob Astor's American Fur Company, it was abandoned in 1822, although federal prohibition against the sale of liquor to Indians continued in effect until well into the twentieth century.

This law against trading or sale of alcohol to tribespeople, like some other laws designed to regulate morality, was not very effective. Vast distances and the isolated nature of the fur trade hindered enforcement. Traders were licensed by officials of the Bureau of Indian Affairs, who could revoke the trade privileges of those dealing in liquor. The frontier merchants often told stories of each other's violations, real or fabricated, to hurt the business of competitors, and some agents cooperated with ruthless traders to exploit the Indians.

Not all sales of distilled spirits in the fur trade were made to Indians. Saint Louis, Missouri, citadel of the trade in the West, was a funnel through which thousands of gallons of whiskey flowed when mountain men, sometimes accompanied by Indian wives, arrived with their beaver furs each summer after the fall and spring hunts. Then in the mid-1820s, Rocky Mountain Fur Company owner William H. Ashley established the rendezvous to beat competition from other fur merchants in Saint Louis. Instead of each mountain man hauling his bundles of furs to far-away Missouri, large numbers of them would assemble at a point such as Pierre's, or Brown's, or Jackson's Hole to congregate with Indian women and barter pelts to shrewd traders from the company.

And barter they did, for whiskey, traps, guns, powder, ammunition, blankets, items of clothing, knives, and other possibles, but above all, for whiskey. Corn juice was a dominant part of these transactions. Liquor was hauled west by pack animals in the form of colorless, "pure" grain alcohol, which was actually only about 95 percent pure. It had been dehydrated to cut transportation costs, but before it was sold or traded it would be cut three, four, five, or even ten times with water. The company would then mix it with red pepper, molasses or brown sugar, and bits of tobacco to give it color before selling it to the thirsty fur men. The drunker the trapper, the greater the dilution!

Long distances and forbidding terrain limited the selling of corn liquor in the Far West. Peter H. Burnett, in propagandizing for Oregon settlers, wrote to the *Niles National Register* in 1844, "No country in the world affords so fair an opportunity to acquire a living as this. I can see no objection to it, except it be by a man who loves liquor, for he can get none here."

In time that condition would change, as it already had in the corridor of the Santa Fe Trail and trade. If Spanish and Mexican customs officials lacked a name for the corn liquor from their neighbor nation to the East, they soon devised one. *Quintoque juizque*, Spanish phonetics for Kentucky whiskey, appeared as an entry on some of the *facturas*, or trade invoices of the *extranjeros*, those foreigners who journeyed in their snakelike caravans over the shimmering trail from Missouri. Corn liquor bridged the gap between two great maize cultures of East and West, reaching the Pacific Ocean at least as early as the 1830s when Tennessee trapper Isaac Graham operated a distillery in northern California. It had played a highly significant role in the crisscrossing of a vast continent, while infesting every corner of the land, whether mining camp, cow town, settler's cabin, or military post.

Saloons in the western mining towns dispensed enormous quantities of corn liquor. In some, probably most, mining towns, the saloon business ranked a respectable third behind mining and banking. In addition to the usual reasons for heavy drinking, boredom, social pressures, frequent use of whiskey as a medium of exchange, and pollution of drinking water with mining debris and sewage caused heavy imbibing in mining camps.

When Union Major Isaac Lynde took his garrison along a hot, dry path of retreat from Fort Fillmore toward Fort Stanton, New Mexico Territory, in July of 1862, he neglected to have his soldiers' water supplies checked. Many of the men had reportedly filled their canteens with whiskey from the dispensary of the abandoned fort. A thirsty disaster awaited them and the entire force was captured by the troops of the Texans' Colonel John R. Baylor. Typical of many expressions about conditions at western forts is the following quote from the Helena *Weekly Independent* about the liquor problem on the frontier of the Far Northwest in the late nineteenth century. "The old black bottle is the greatest enemy of the soldier. It will find its way inside the post in spite of all the precautions that may be devised. The ingenuity shown by the men in getting whiskey would make eternal fortunes if turned to another line."

Meanwhile, back at the still the processes of milling and moonshine making, of tax dodging and bootlegger blockade running, as well as legal distilling, repeated themselves. Generally, the producers were called moonshiners and the transporters bootleggers, but these terms were often used interchangeably. Distillers made the liquid product with almost biological exactness, generation after generation, in settlement after new settlement. From Great Lakes to Gulf of Mexico and from Appalachia west to the cutting edge of the farm frontier, every wooded section of the wide corn country had its Still House Holler, its Moonshine Cove, or Whiskey Gulch. Frequently, as noted in an earlier chapter, mill and still were combined at one location, since the corn-grinding process was a necessary part of both. One western area during the 1830s had three stills within a three-mile stretch.

It was as routine as a deer going to a salt lick or a watering hole. Nearly every pioneer family that did not make its own home brew would send a family member, often a child, with a sack of corn to exchange for whiskey at the nearest still. Varying slightly with the strength of the liquor, a gallon for a bushel of corn was the usual rate of barter. Somehow, whether the family was rich or poor, whether corn crops were good or bad, they always had a jug of whiskey on hand, and more than one if there was a husking bee or house-raising scheduled. Inevitability (or desirability) seems to have given whiskey respectability. Except for paupers, there was little or no disgrace or reproach for imbibing, even among the western clergymen, who found their justifications in that oft-quoted biblical passage, "Drink a little wine for the stomach's sake." The term *a little* was not well defined, and most adults consumed whiskey, the "drink of the West" with regularity. From 1790 to the 1830s, Americans annually drank whiskey at the rate of about five gallons per capita, in contrast to just over two gallons today. By the 1830s the temperance movement began to cut alcohol consumption sharply.

"Fuller'n a fall corn crib" was a phrase reserved for those who had overindulged. Women customarily drank whiskey in the form of toddy—the distilled spirits blended with hot water, a sweetener (such as honey, maple sugar, or "store" sugar), and spice when available. Children, too, drank toddy, although there was some uneasiness about this. One wonders if the concern was more for the welfare of the youngsters or the level of "liquid dynamite" in the jug. Little wonder, with such early exposure, that America was a nation of drinkers. John Adams despaired at the fact that it was the world's most intemperate country. Just as it was considered masculine and healthful to drink, it was considered impolite or even dangerous to refuse a drink.

Keeping the millions of jugs filled in pioneer America, and the hundreds of thousands of barrels for trade to distant points was no small order. Between 1820 and 1830, seventy-five thousand barrels of whiskey were shipped from western states down the Ohio River to New Orleans. Before the 1830s there was not a solitary millionaire or multinational corporation to exploit the business, merely thousands of small rural entrepreneurs—mill operators and farmers. The requirements and capital outlay were not great: a copper still and "crane's neck," twenty feet or so of coiled copper tubing, a few tubs and barrels, access to a grain mill and a stream of fresh water, and in many instances, a knack for the art of concealment. Needless to say, there was another essential—corn.

The early processes, during the seventeenth and eighteenth centuries, were simple but basic to the more modern techniques. About one-fifth to one-sixth of the corn to be used in a batch, or run, was moistened and kept warm until it sprouted. Germination of this malting grain created diastase, which converted starch to sugar. After the remaining four-fifths of the corn had been ground fine, cooked as mash in a copper still, and placed in the necessary number of tubs, the malting or germinated corn was ground and added in equal proportions to each tub of mash. The malt was the starter which caused fermenting, or working, of the mash. Yeast, often a scarce commodity, was stirred in when available, and some producers added sugar to speed the fermenting, and a few measures of rye. After three to five days, with daily checking and stirring, the liquid mash fermented to beer and was ready to distill. Fermentation had converted the starch to sugar, which then turned to alcohol, amounting to perhaps ten percent of the mash, or twenty proof. A buyer could always test to see if whiskey was at least a hundred proof since fifty percent was the lowest alcohol content which would enable the mixture to burn. Other proof tests included shaking to check the bead and a test wherein the rate at which a particle of tallow sank in the whiskey was an index of its proof.

In order to extract the alcohol in a purer form, the distiller cooked the mash in a copper still. This large kettle, often hammered from a copper sheet by the ingenious moonshiner, was encased in a stone and clay furnace so that heat from the fire at the bottom of the furnace circulated around the still. Corn cobs or pine knots were used to kindle the fire of logs shoved or fed into the firebox at the bottom. The container was protected from direct flame by a flue in order to prevent sticking of mash during cooking. Since alcohol had a lower boiling point than water, its vapors rose to the head of the still, passed through the crane's neck, and condensed in the worm of coiled copper tubing set in a worm tub, or barrel of cold water. The tube had been coiled by bending around a stump. From this worm, the condensed alcohol flowed into a can or jug receiver, and the run was finished. Double and triple distilling produced a smooth but deadly whiskey known as double back, or twisted whiskey. This process could remove a higher percentage of water leaving up to about 95 percent pure alcohol, or 190 proof. Sometimes still operators dumped the waste, or distillers' "wash" into a stream, but there were objections to pollution even then especially since revenuers could find a still very readily by detecting evidence of mash in a creek or river. Some revenuers reportedly rode horses through moonshine country. If a horse refused to drink from a creek, the official would investigate upstream to see if there was a still with mash dumped into the water. Distillers found it much more profitable to feed the used mash to hogs. A farmers' almanac of 1823 reported that the "wash" fattened livestock much better than whole grain. This saving cut the cost of whiskey production noticeably.

Distillers made many improvements during the nineteenth century, using steam instead of direct heat, devising revolving rakes (to stir the mash in the still and prevent burning), fractionating, and continuous distillation. Some changes resulted from accidental discoveries. Freshly distilled whiskey had a "new" sharp taste which was removed by aging in wood barrels, preferably white oak, for two to eight years. During the mid–nineteenth century a shortage of whiskey barrels in some areas made it necessary to use containers in which salted fish had been shipped. One distiller, in an attempt to get rid of the fish flavor, burned the insides of the barrels before filling them with whiskey.

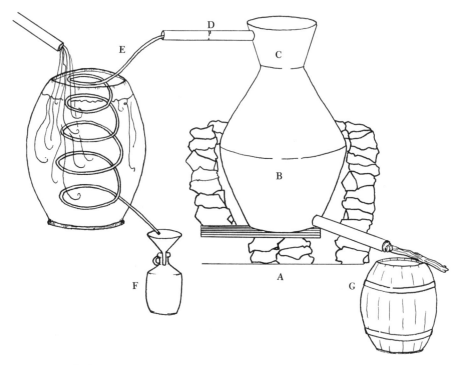

"Watch them jugs a-fillin'." In this cross-section of the distillery, or still, mechanism, the furnace (A) heats mash in a copper kettle (B). Alcohol vapor rises to the cap (C), passes through the cap arm (D), and then circulates through a coiled copper worm (E), where it is condensed to liquid by cooling in the water barrel. It then drains into the jug (F). After the batch is run, residue from the kettle is drained into the slop container (G) for feeding to livestock.

These charcoal containers quickly gave the whiskey a superior flavor and revolutionized the aging process. Charcoal containers are now required by law for mellowing and aging whiskey. (Another account holds that the charcoal process was an accidental discovery by Baptist minister Elijah Craig of Kentucky in 1789.)

There were other factors that had effects little short of revolutionary. Whiskey distilling in the Confederacy suffered a severe blow, from which it took years to recover, when Union forces captured Tennessee copper mines during the Civil War. The South requisitioned copper kettles and condensing worms from stills in order to make the indispensable copper percussion caps for caplock muskets. Confederate leaders also strove to curb whiskey production in order to conserve

A. W. Thompson drew the SOUTHERN MODE OF MAKING WHISKEY for HARPER'S WEEKLY, December 7, 1867. Countless hundreds of such covert distilling operations dotted the backwoods.

grain for food in that time of critical shortages. What better proof that the South was dedicated to winning the war!

Renewal of federal taxation, first briefly during the War of 1812, then continuously from the Civil War to 1920, and the National Prohibition (Eighteenth) Amendment, which was ratified in 1919 and took effect one year later, had far-reaching effects. The constitutional regulation of 1919 prohibited the manufacture, sale, transportation, importation, or exportation of intoxicating liquors for beverage purposes. It signaled the return of small moonshiners to a business which had been falling more and more into the hands of large corporations. Some estimates indicate that per capita consumption of distilled spirits may have increased slightly during the Prohibition era.

Modern blockade running and pursuits were carried on with automobiles instead of horses. Some moonshiners set up their stills in "dry hollers," piping water underground to prevent revenuers from ferreting out their secret operations by simply following streams into backwoods areas. Others hid their operations by setting up in the open where they might be least suspected to house their stills. In 1956 when the English agriculturalist Hugh Willoughby visited eastern Kentucky, a land that had "contracted out of history and stayed put," he reported that warning shots were fired to frighten a man away from a still. I recall living between two such bootleg joints, each less than a quarter of a mile away. One was in a small shack in an open field beside a major highway. The other, not discovered until the house was sold years later, was set up underneath the kitchen and reached by a trap door in the floor, which was covered by linoleum. Harry H. Meek, the great-grandfather of this book's illustrator, was a Memphis railroad man who had a reputation for being able to "sniff out" concealed whiskey in incoming cartons. He reportedly confiscated it with safety, since illegal importers could not lay claim to it.

With the repeal of the Prohibition Amendment in 1933, small operators were greatly curtailed, leaving the liquor-making field mainly to big corporations. Modern medicine contributed to the decline by gradually undermining the myth of white moonshine's healing powers. Thus the sporting age of "mountain dew," "white lightnin'," and "rot gut" distilling in the corn and timber country was shut down to a mere trickle compared to its bountiful gush in years gone by. "Fast buck" artists further reduced the business by using cattle feed for mash, brewing in the radiators of automobiles, and adding potash to give the liquor a bead—that is, bubbles that form when freshly made whiskey is shaken. In a few states—Kentucky, North Carolina, Tennessee, and especially Georgia—the old craft has been kept barely alive. Federal revenue officers seized 481 stills in 1977. During 1978 the number of seizures dropped to a new low of 361.

Though less dominant than its role as a fount of potent drinks, corn's contribution to smoking has been noteworthy. Indians had long smoked dried corn silks instead of tobacco and often mixed the two in their clay and stone pipes. Their inquisitive white visitors mimicked the practice and soon improvised with other parts of the corn plant. Husks became cigarette wrappers for smoking both tobacco and corn silks. Many a youngster, even into modern decades, experienced that first smoke behind the barn or corncrib with a cigarette of corn shuck wrapped around brown, dried corn-silk "tobacco."

Pioneer settlers, probably before the 1820s, came up with another variation on the smoking theme, the corncob pipe. They would age cobs and cut them into blanks (two or three from each cob). The soft cob centers were removed, and inside and out, the pieces were trimmed and burnished down to the hard, woodlike barrel or cylinder. Bottom plugs were fixed in the bowls, and they were ready for drilling stem holes and inserting stems. Hollow plant stems from wild honeysuckle, elderberry, or the easily replaceable dried goldenrod served well as pipe stems. With a minimum of effort and time, the pioneers had devised the simplest smoking toys imaginable from those all-purpose oddities of nature, corncobs. If a smoker lost, broke, or

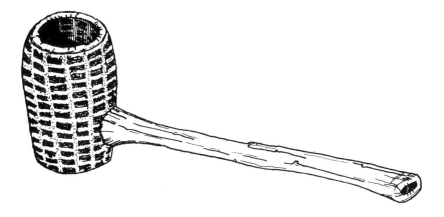

Corncob pipes, commonly smoked by both men and women, were easy to make and simple to replace.

burned out a pipe, another "barnyard briar" could be prepared and puffing smoke within minutes. The cob, preshaped by nature, was simpler and less expensive to work with than clay, stone, or hardwood.

What percentage of pioneer settlers smoked corncob pipes is impossible to determine, but it was probably considerable among minors and high among adults. Before the Civil War era, it was common for both men and women of all ages in frontier regions to smoke pipes of cobs or clay.

Despite the simplicity and ease of home-crafting corncob pipes, that business went commercial by 1869. The Missouri Meerschaum Company, which took its name from the meerschaum mineral content in pipe clays, opened a plant at Washington, Missouri. Eventually, it worked out mass-production techniques, using bone mouth bits (now plastic), wood stems, and for pipe bowls, cobs aged up to seven years. The Meerschaum Company was not a large business, and by no means did it wipe out home-made corncob pipes, but its production was large enough to give it and the Hirschl and Bendheim Company of Saint Louis a near monopoly over commercial manufacture.

Down the pathway of the centuries since the arrival of Europeans, corn has kept a preeminent position in whiskey making almost as convincingly as it has maintained leadership of food and feed crops. The result, in both early and modern times, has been a mixed benefit for Old World settlers and their descendents. Alcoholism, long a serious problem, has been increasing at an alarming rate in recent years. Disruptive as corn liquor was to the Indians, who gave this and other crops to their trans-Atlantic visitors, there are those who observe that maize in its liquid form and smoking tobacco have become the Indians' revenge for the loss of their lands and culture at the hands of the invaders.

From recreation and food to drink and smoke, the settlers' uses of Indian corn spanned a wide spectrum. But before pioneer America had satisfied its appetite for adaptation of the phenomenal maize plant, the list of nonfood uses would lengthen like afternoon shadows, far outdistancing all other plants of the era.

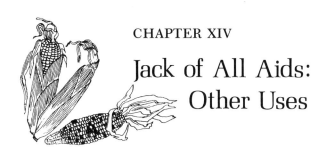

CHAPTER XIV

Jack of All Aids: Other Uses

> There is no plant or vegetable cultivated . . . that is capable of being applied to so many purposes of utility as Indian corn.
>
> EDWARD ENFIELD, 1866

If maize or corn seemed heaven-sent to the Indians for its role as staff of life in their diet, it must have appeared nothing less than a divine plant to the pioneers. Not only was it the first food crop and basic to the nutritional needs of themselves and their livestock and poultry, but it met scores of other requirements in their daily lives. In its by-products, corn was a "jack of all aids," adapted to more nondiet purposes than all the other pioneer farm crops combined. Today, scarcely noticed by most Americans, there are more than five hundred industrial uses of corn. Pre–twentieth-century farmers were keenly aware of every use of each part of this versatile plant.

Owing to the variety of its parts, the ear offered the greatest range of uses in early times. And of this, the cob, called the core in early colonial times, was applied to the largest number of nonfood needs. It is a structural oddity, a marvel of nature: a light, round stick of woody fiber with a pithy core. Not surprisingly, its farm and frontier uses were largely as wood, mainly firewood.

Month in and month out, year after year, that great majority of Americans who lived on farms cooked their meals and took the chill out of their cabins with wood fires in open fireplaces or, increasingly after 1840, iron stoves. The work cycle was similar all across the land. Late on a snowy evening at nearly every farm, a member of the family, often a youngster or the woman of the household could be seen, jute bag or gunny-sack in hand, crunching through the earth's white crust to the corncrib. In a few minutes the figure, silhouetted sharply against snowdrifts in the gathering gloom, returned with a bulky but not-too-heavy sack of cobs on one shoulder. So important were they for fuel that cobs were even raked out of muddy pig pens, dried, and used for heating and cooking.

Back at the dwelling the cobs were poured into a bin or left in the bag, ready for starting the morning's fire. Although corncobs burned too quickly to replace hardwood in providing sustained heat and good coals, they were the most widely used kindling wood of rural America. They could provide about one-third the BTUs of dry hardwood by weight, and they were far superior to the twisted faggots of straw and the buffalo and cow chips burned for fuel on the plains. In cool and cold months, at least one fire in the farm cabin was kept alive at all times. Each night, cabin dwellers banked their live coals, that is, buried them with ash. In the morning they would scrape aside the gray ashes and carry coals, which were still red hot, in a shovel or metal firebox to other fireplaces or stoves as needed for the larger houses. Sometimes neighbors would borrow live coals from one another if a fire had gone out. Corncobs were placed on these coals, and sticks of wood piled on top. In moments there would be a roaring, crackling blaze. At times too, if cobs were plentiful, pioneers heaped them into the fireplace to take the chill off the room quickly or to provide light. Corncobs produced less smoke than wood and little odor. They were better than wood for heating a quick lunch or a pot of coffee.

When there was a surplus of grain and prices were low, entire ears as well as cobs were burned. During the Civil War, when corn shipments from North to South came to a standstill, many northern farmers burned their surplus corn—grain, cob, stalk, and all—as fuel. On the treeless prairies and plains corn was grown to fill both the crib and the woodshed. Settlers on the plains in the 1870s were known to burn their grain as a protest against discriminatory railroad rates. Outdated seed corn mixed with coal is presently being burned at Logansport, Indiana, to generate electric power.

Not all cob combustion took place in stoves or fireplaces. In the absence of wood, firemen sometimes dumped corncobs into locomotive fireboxes to heat the boilers and generate steam for motive power. In preference to wood, many farmers used them to smoke meat. Cobs also made good torches or brands. Farmers soaked them in pine pitch, hot oil, or grease, impaled them on rods, and lighted them. These torches would burn for many minutes, giving off far more light than lamps, candles, or lanterns. Lighted torches were quickly drawn along trails of ants, thereby providing the best control of these pests before the era of modern chemical repellents and killers. Some crude lamps

Corn meets corn in this whiskey jug plugged with a corncob. Most large jug and bottle stoppers were made of cobs.

with corncob wicks, resembling betty lamps and crusie lamps, were home-fashioned or made by tin smiths, and pith from cobs was soaked in oil and placed in metal pincers of rush lamps when rushes were not available.

Some of the noncombustion purposes of corncobs have been discussed. They were used as corn shellers, mousehole and knothole plugs, fishing corks and hook holders, missiles for corncob fights, and as a base for producing a substitute for baking powder. Practically everywhere cobs were used as jug and bottle stoppers. William Cobbett wrote in 1828, "Never did I see any other corks in a farm house in America." The taper, roundness, and slightly cushiony surface made cobs ideal for stoppers. And as back scratchers, these fibrous cylinders were without equal, for they were soft enough to prevent irritation, rough enough to scratch. Some skin ailments were treated by rubbing with medicinal wands consisting of cobs soaked in solutions of grease and herbs. Such a massage in Ozark country was called a "plumb

Dried cobs made excellent handles for tools like this file.

good cobbin'." Given their unique form and structure, it is not surprising that cobs were the most widely used scrubbing brushes for cleaning pots, pans, and plates.

There were other sanitary and not-so-sanitary applications of the corn cob. From the Indians, pioneer settlers learned to use it as tubular toilet tissue, which was subsequently burned. Outdoor "chick-sales," or privies, often had handy bins for these wipers. Fortunately, although woody, the cob had no splinters.

For the most part, the nonflammable ways of employing corncobs were associated with their sticklike structure. The soft centers made them instantly adaptable as handles for files, chisels, awls, and other small tools. If a handle was broken or missing, the user could snap off a cob to half or two-thirds length and quickly press it onto the tang of the tool. Some handles of this type lasted for years. Similarly, one could force cob sections on the sharp points of scratch awls, punches, and ice picks as a safety measure while these tools were not in use.

The list of uses for cobs was almost as wide as the creative instincts of the pioneer era. In crushed form they served as fertilizer, livestock feed roughage, absorbent for removing solvents applied in the cleaning of furs, mulching material and soil loosener for garden plants, and litter in poultry pens.

One of the greatest uses of whole cobs was as barnyard litter for livestock. The teeth and hooves of horses, cattle, hogs, and sheep cropped the herbs and chopped the earth of barnyards so that, in seasons of rain or melting snow, nearly every such yard was a quagmire. Farm hands scoop-shoveled large quantities of cobs into the mud to make walking less soggy for beast and human.

As if to balance the rough outdoor and work-related uses, there were also adaptations of corn cobs for delicate indoor needs. The lyric line "Corn cobs twist your hair," from a song of the 1830s, indicates that pioneer women may have adapted these light, rough-surfaced cylinders as hair curlers. Short sections of cobs with nails inserted lengthways through the soft centers were serviceable as cool hand-holders or skewers for eating hot roasting ears of corn. These were just as effective as modern ear holders and were much cheaper. The coloring from red cobs was used to tint drinks a delicate pink shade and is believed to have served as a dye for fabrics. Cobs with the pithy centers punched or burned out (with a red hot rod of metal) and plugs inserted in the ends functioned in countless frontier households as containers and shakers for seasonings and as unbreakable bottles for powders, pills, and other small, dry material. In the last third of the nineteenth century, "corn cob down," the chaff left over from manufacturing corncob pipes and from cleaning shelled corn with fans (called vans in earlier times), was used to stuff pillows, mattresses, and the like. The list of the ways in which pioneers employed the versatile cores is almost enough to make one forget that their primary purpose was to support the no-less remarkable grains in growth and storage.

Added to the corn kernels' diverse use for food and drink was a sizable range of nondietary adaptations. The multiple recreational uses were almost as well known as the recipes. As seen earlier, the grains served in various games, as decorative art, and for fish bait. Corn oil, extracted at the distillery, "burned brilliantly" in the lamps of the early 1800s. By 1860 it was used to illuminate lighthouses on the Great Lakes and other western waters. One common adaptation of the whole dried ear with husk removed was as a darning egg inserted in stockings while holes were darned, or mended with a simple crossweave.

Perhaps the grain's most important nonfood use was as money. Indians of the Americas had long counted corn kernels and ears just as they had cacao beans, cowry shells, beads, bird feathers, arrow points, and animal pelts for their economic transactions. In their specie-short society, pioneer Americans, far more often than not, paid their bills in kind—that is, in the things they produced. Corn was coin throughout much of America, whether in the toll meal that the miller kept as his share for grinding or in the portion of a field's production that the sharecropper turned over to the landowner or in corn whiskey measured out to pay the neighborhood preacher and circuit rider as well as the teacher. Thomas Jefferson bought a mill seat (site) with corn as money and used this grain as payment on countless other occasions. "Corn is always a ready money article," he wrote. "This I cannot sacrifice for any considerations." Sam Houston, schoolmaster at Marysville, Tennessee, in 1811, received eight dollars per pupil each term in corn, cotton goods, and cash. One hundred gallons of corn liquor were used to pay part of the minister's salary at the First Presbyterian Church of Cincinnati, and a newspaper in Ohio specified the quantities of corn products—corn, whiskey, pork, and beef—which it would accept as payment for subscriptions. Down on the Texas plains in 1823, American settlers were each expected to give Stephen Austin ten to twenty bushels of corn to pay the expenses of Don Erasmo Seguin, deputy to the Mexican Congress. Spinsters, weavers, and washerwomen accepted cornmeal as payment for their work.

On October 8, 1631, Massachusetts Bay Colony made corn legal tender for all transactions unless coin or beaver pelts were specified in an agreement. Taxes as early as the 1640s and rents and marriage license fees were frequently paid in corn. The colonists resorted to corn as money in the chronic cash shortages because it was widespread, well known, usable, and liked in every colony (and later in every state). Its value, like that of cowhides in California, beaver pelts throughout the West, and rum in the coastal towns, could be easily agreed upon—more easily, in fact, than some of the clipped and alloyed coins of early days.

Country stores from colonial times to the post–Civil War era were more than merchandising houses. In effect they were also banks, extending credit to settlers from times of need to harvest seasons when the farmers could hope to have enough surplus crops to pay off their debts. Depending on the area, when debtors paid the store proprietors, often as not it was in corn or corn products such as salt pork and whiskey. Prices were usually 20 to 30 percent higher for credit than cash transactions. Still another function of the country store was its service as a clearinghouse for barter deals or payments in kind, whether involving immediate barter payment or credit. Corn's high-bulk–low-value ratio limited its direct use as money to relatively short-haul situations, but when it was converted to meat (either salted or on the hoof), butter, or whiskey, farmers and merchants could profitably ship it over great distances.

Corn, particularly as corn whiskey, was the most common ingredient in home remedies for whatever ailed pioneer Americans. Except in Gold Rush California, where they were "plenty as blackberries," according to one physician, medical doctors were usually scarce in frontier areas. Because of the nature of medical practice during pioneer times, this was perhaps fortunate. Causes of disease were unknown, and treatments were determined by symptoms and superstitions. Patients, if they survived, did so, often as not, in spite of the treatments—or mistreatments. Home remedies, which abounded in the thousands, frequently consisted of blends of superstition and whatever strong-tasting and pungent-smelling potions were at hand, including vinegar, tobacco, kerosene, salt, sugar, honey, tea from acrid or distinctive-tasting roots and leaves, turpentine, kraut juice, tar, pitch, quinine, onions, red pepper, brandy, wine, and of course, whiskey. The sick often lived and improved, probably very largely as a result of psychological or placebo effects as well as nature's healing processes.

Distilled corn liquor, the most universal medication of pioneer settlers, was used in so many mixtures for so many illnesses as to defy description. Thus in home medicine as in many other ways, corn emerges as the principal American crop or plant. Other grains, particularly rye, were important in whiskey production, but they were not as significant as "corn juice."

Whiskey was used externally, rubbed on as a liniment for headaches or sore muscles and applied as an antiseptic—and here, because of the high alcohol content, the amateur physicians were on solid scientific ground. It was in the internal applications where use became abuse. Frontier healers considered whiskey to be almost a cure-all. Usually they mixed it with honey, sugar, or candy, and they used it to treat coughs, colds, croup, whooping cough, sore throat, colic, dysentery, consumption, pneumonia, asthma, and arthritis. Various kinds of roots, leaves, and bark were steeped in the beverage to deal with colic, pneumonia, rheumatism, hookworm, catarrh, snakebite, "yaller janders" (jaundice), gall bladder, and many other ailments. For toothache, holding whiskey in the mouth was a standard treatment. The supply was replenished from time to time as the patient swallowed "involuntarily." To waste the treating fluid would have been unthinkable! For frostbite or snakebite, for broken bones or breaking a fever, for stress or fits of depression, most settlers turned first to the jug for relief and treatment. Jamaica ginger, an alcohol-based patent medicine, was used for relieving congested bowels. Among pioneer pain

remedies, whiskey ranked very high even after the advances in anes-
thesia during the Civil War. Its most potent pain-killing use was in lau-
danum, a mixture of alcohol and opium.

Beyond the specific prescriptions of whiskey in some form, there
was a good-for-whatever-ails-you attitude toward the brew. No doubt
this was an excuse for frequent indulgence, if not overindulgence. Not
that excuses were necessary. Had corn whiskey been a good preven-
tive for illness, the frontier should have been the healthiest place in all
America. In fact, the nation as a whole should have been healthy,
since, as noted previously, whiskey consumption in early times was
much heavier than it is today.

Related to the use of corn liquor as a cure was the taking of whiskey
for warmth. It was typical of the frontiering populations that they
counted on the fireplace for external and firewater for internal control
over frigid temperatures. Modern science knows that they were fool-
ing rather than fueling themselves with that burning sensation in the
stomach.

Corn grains had medicinal uses in forms other than "liquid dynamite." There were superstitions about getting rid of warts and boils by
cutting, putting a drop of the blood on a corn kernel, and feeding it to a
chicken to make the ailment or blemish disappear. The *New England
Farmer* of 1829 recommended corn oil, which at that time was separated through the distilling process at the rate of two quarts per bushel
of grain, as a good medicine if taken like castor oil. Settlers used cornmeal poultices almost as widely as liquor. Sometimes they simply
moistened the finely ground grain with boiling water. More often they
mixed it with milk, animal fat, and other substances (such as eggs,
salt, onions, turpentine, and herbs), heated it, and applied it in cloth or
layers of corn shucks to the afflicted area. Such poultices were utilized
for remedying toothaches, boils, sprains, and other conditions involving swelling or inflammation. The treatment had some scientific valid-
ity because it generated moist heat. An old Ohio remedy called for
eating large amounts of cornmeal to reduce stomach and intestinal
damage from the swallowing of fish bones, ground glass, or other
abrasive matter.

Many emigrants packed valuables such as china or glassware in barrels of cornmeal to prevent breakage in transit to their new homes.
Ground corn in another form was widely used by settlers. It provided
the cheapest and most readily available starch, not only in foods but as a
garment stiffener, as sizing for cotton thread, as thickening substances

in calico printing, and as a nonactive ingredient in pills—nothing like the countless uses today, but important nevertheless. Women made starch by pounding the grain, preferably flint corn, steeping it for several days, and boiling it down, then drying it to powder. In the early 1840s, Thomas Kingsford greatly expanded cornstarch consumption when he opened his commercial production plants, first in New Jersey, then at Oswego, New York. Today the Kingsford Company manufactures cornstarch in Connecticut.

Rancid or foul-smelling wooden barrels and casks, butter tubs, buckets, and the like were a problem for pioneer settlers. These were difficult to clean since the odorous substances penetrated the wood staves. Farmers' almanacs of the early nineteenth century told how to avoid the cost of replacing these musty, smelly containers. Fill them with cornmeal and water and let the mix stand until it had thoroughly fermented. The results were a clean-smelling keg and a quantity of mash, which did double duty as excellent hog feed.

Fermented corn had other uses, some of which have been largely neglected by recorded history. Pure corn alcohol served as a lamp fluid and laboratory burner liquid in the nineteenth century. Various industrial and commercial functions developed. One that was of particular value was the use of grain alcohol as a baffle, antioxidant, and antifreeze for protecting valuable instruments. John C. Fremont carried it to preserve his topographical and surveying instruments on the far western frontier during the 1840s. It reduced breakage of delicate instruments from the jarring of rough terrain, and it would not freeze in the coldest mountain temperatures, although at around one hundred degrees Fahrenheit it suffered some evaporation loss. The greatest "evaporation" came from sly tippling by thirsty mountaineers on the expeditions. This loss was greatly reduced by adding a solution of tartar emetic, which caused instant vomiting when taken internally. But the trails were so rough that most instruments were eventually broken despite precautions.

Corn husks, unlike the grain, had no direct human food value, but they more than compensated for this with their versatility, including their food-related uses. Careful removal of the mature ear left the dry husks in a hollow tubular form which substituted for animal intestine as a sausage casing. The paperlike inner corn shucks, commonly rolled as cigarette wrappers, were much more extensively used to wrap tamales, to protect corn cakes for cooking in ashes, and to "put up" cakes of sugar. They were packed over melons and pumpkins and

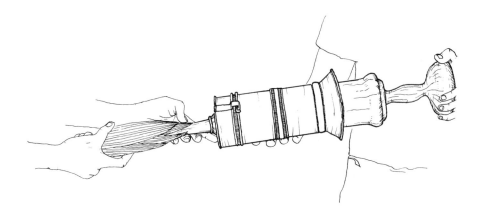

Corn-fed pork sausage was pressed into a husk wrapper using a tin cylinder and a wooden plunger.

wrapped around apples, pears, and other fruit much as paper has been used in modern times.

Given the scarcity of paper in Early America, it was inevitable that corn shucks would be utilized for writing. Charcoal, graphite, and ink made of dyestuffs such as nut hulls, indigo, oak balls, and soot were used with quill pens to write on the husks, as was a black ink made from the dusty inside of corn-smut pods. The *American Farmer* magazine of 1828 reported that Burgess Allison and John Hawkins had patented a process for making paper from corn husks in 1802. A quarter of a century later, Nicholas Sprague of Fredonia, New York, improved this process, making good quality, raglike paper from husks.

Shucks from the corn plant, when dry, had a resilient, cushiony property which made them adaptable to a wide range of shock-absorbing needs and situations. Braided, or stuffed into leather tubes, they made good horse collars; and shredded, as chair and couch upholstering and mattress and pillow padding, they were essential to the sleep of many settlers. As early as 1820, American magazines reported the great success of corn-husk padding. Far better than straw, it lasted five to ten years, did not scratch the sleeper, and seldom developed lumps. And it harbored fewer of "those creeping or skipping things" that bit and sucked blood. Abraham Lincoln was born in Kentucky on a bed of corn shucks and bearskins. A Tennessean stated in 1849 that corn-shuck mattresses were better than dried moss.

An old southern song suggests that shucks were useful for making some of the dwellings that housed the mattresses. "On Tom-Big-Bee River" refers to a hut made of husks. Artisans braided husks into chair backs and seats and wove them into mats, cords, baskets, hats, and sometimes curtains and shoes. Insoles of these leafy sheaths were placed in wet shoes to make them more comfortable. Shredded and tied to the end of a stick in bundles, shucks made an adequate cleaning implement that, in appearance and application, was a combination of mop, broom, and feather duster. Husk dolls, flowers, and other ornaments were long a part of the American household scene.

The long, pollen-catching, threadlike silks of corn were a natural for hair on the heads of husk dolls, but as smoking tobacco they were less appealing than what Sir Walter Raleigh discovered and took back to England from America. As with whiskey and cornmeal poultices, silks from the bountiful crop were relied upon to cure illnesses. They were soaked or steeped in water, which was prescribed as a remedy for urogenital disorders, dysentery, and other ailments. Interestingly, modern drug manufacturers rely extensively upon corn, using silks in a variety of ways and producing penicillin and other antibiotics from the steep-water of the grain.

Not to be outdone by its fruit, the cornstalk was caught up in a host of farm and frontier uses. Like the husk, it was fashioned into human-like figures—not dolls, but scarecrow manikins to stand in the fields of newly planted corn to frighten away the winged enemies. Another use of stalks was as light fences or walls. The sturdy, bamboolike stems, ranging from one to two inches in diameter at the base, sharp from the angle of cutting, were pressed into the soft earth in very close formation. Reinforced with horizontal sticks, wire, or rawhide thongs, the stalks provided protection for gardens against wind or small mammals and as fences to contain chickens, ducks, geese, and turkeys. The wing feathers of birds were clipped to prevent flight over fences.

Human protection, too, was afforded by cornstalks. In the construction of roofs, particularly out on the sodhouse frontier, builders packed dried stalks tightly together atop the gently sloping split-log rafters. They placed sod in two layers, each about two and one-half inches thick, on top of the corn. The bed of stalks greatly reduced the amount of dirt that fell into living quarters below.

Corn-shucking time, with its shivery dawns and shortening days, was a reminder to every log-cabin farm family of the central and northern latitudes that howling gales and driving snows would soon lash their dwellings. Cornstalks, the dwellers found, could keep the

Settlers would bank their cabins with fodder from the corn harvest to insulate them against winter's cold. Snow would finish the protective blanket.

cold as well as the dirt out of primitive shanties. "Bank up the house," the almanac warned its readers. In late October or early November, wagons laden with shucked corn and sheafed fodder creaked past the farm cabins on the way to barns and corncribs. Bundle after bundle, the workers tossed fodder off the wagons, then packed the dry stalks lengthwise along the outsides of the cabin walls in sloping piles up to the window sills and sometimes higher. They propped heavy lengths of wood over the slanting fills of corn to prevent blow-away in wind storms. In time, snowdrifts would hold these piles in place and give a strange pyramid appearance to the log houses. Thanks to the fine insulating properties of Indian corn, American cabins, from the Blakes' of Connecticut to the Garlands' of Nebraska, would be literally "tucked in" for a much warmer winter whether or not their chinking was tight.

Inside the cabin, an occasional farm family would use cornstalks for lighting and heating along with cobs. The fibrous pith of these heavy stems was occasionally used as lamp wicking, and there were instances when cornstalk partitions provided all the privacy that was available in the primitive, one-room shanties. Earth floors, always characteristic of newly built frontier farm cabins, were cool in summer but cold in winter. A layer of cornstalks, perhaps blended with hay, was scattered on the floor by some settlers during extremely cold snaps. This rough organic carpet would crackle noisily underfoot as the inhabitants moved about the dwelling, but noisy steps on some of nature's best insulation were better than cold feet on bare earth.

Outside at the barn lot, livestock too trampled noisily on corn fodder which, along with the cobs, had been scattered as litter. Some settlers, if they had the materials and the time, insulated their barns and sheds, either by banking or with stalk stacks against interior walls as protection for both animals and human workers during frigid months. And there were countless instances in the shock-studded fields of winter when the carefully stacked shocks of corn provided relief from the cold or protection against death from freezing. They were common resting places for hobos and other wanderers, just as they were hideaways for fugitives. There were times, too, when a rural partygoer tried vainly to find his way home late on a Saturday night through falling snow and corn-liquor haze, and had to settle for the warm shelter of a shock nest to sleep it off.

Wanderers were not alone in their discovery that corn fodder offered shelter from the elements. Southern back-country people in the early 1700s were known to live in roofless stalls for want of nails to complete their dwellings. At night and during rainy and cold weather they retreated to nearby fodder piles or stacks.

As winter's ice and snow were followed by rain of spring and heat of summer, pioneer settlers found other uses for the leftover dried corn plants. They packed bundles of fodder into gullies to stop erosion and built ridges of earth-covered stalks at a slight diagonal across downhill wagon routes to prevent runoff from following wagon ruts, building momentum, and washing out entire roads. These ridges also served as rests, called "thank you ma'ams," to prevent wagons from rolling downhill when drivers stopped their teams to rest. Just as bamboo was used in other times and climes, rural folk of the corn country placed stalks on top of arbors to provide shade in livestock and poultry pens, on porches, and for roadside stands. Dried stalks insulated rural icehouses where ice from streams and ponds preserved pioneer food supplies well into summer. Pest-plagued farmers would place faggots of fodder in continuous line along the edges of fields and later burn them to destroy the insects that had concentrated there for shelter.

As the rural nineteenth century wore on toward the urban and technological twentieth, the list of corn uses was extended. Stalks were made into varying grades of paper. The great American battleship building boom of the 1880s and 1890s was influenced by cornstalks, as shipbuilders packed the stalks into coffer dams to swell and instantly seal off water between walls in the event they were pierced by enemy shells. Corn oil was used for making paint, varnish, salves, and creams, synthetic rubber, linoleum, and a variety of soaps. Other corn derivatives became important in the manufacture of ink and bases for many types of powders. It was clear that the crude direct pioneer uses were giving way to more refined developments.

Modern scientific advances and industrial techniques have rendered almost completely obsolete the many rural nonfood uses of cobs, kernels, silks, shucks, and stalks, and replaced these rustic, all-purpose materials with plastic, metal, glass, and other substances. Corn products are no longer readily recognizable in such things as panels of fiberboard and nylon panty hose, furfarol for plastics and solvents, and ethanol, a form of alcohol. As we shall see elsewhere in this volume, the present worldwide crisis in energy may be propelling corn into a vital new role. Ethanol is being produced in increasing amounts for use with gasoline (as gasohol) for automotive power. These unique components of the corn plant, once so essential to nondietary needs and to the very lives of settlers, are now either mechanically or chemically transformed beyond recognition or chopped and scattered upon the soil by combine harvesters. So thoroughly have these changes taken place that the old ways of using the corn plant are now a chapter completely closed.

Considered with the countless food and drink benefits of corn, the nonfood uses reflected a high degree of pioneer energy, adaptiveness, and ingenuity in the unending quest for freedom from want. If there was any one crop in America that seemed to spell freedom from want, that crop was Indian corn.

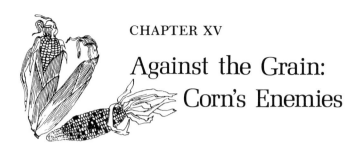

CHAPTER XV

Against the Grain: Corn's Enemies

But when the tender germ begins to shoot,
And the green spire declares the sprouting root
Then guard your nursling from each greedy foe,
The insidious worm, the all-devouring crow.
JOEL BARLOW, "The Hasty Pudding," 1793

The battle for freedom from want applied to beast and bird and plant as well as to human. The battle for freedom from want was a driving force in creatures wild as well as domestic, large as well as small. Clangoring cranes, cawing crows, squeaking mice, and buzzing insects must have concluded that corn was planted and nurtured for their special tastes and needs, that it was given in exchange for the ruthless wrecking of their wooded sanctuary, for the taking of lands to which they had some prior title. Certain it was that corn would be caught in a bitter survival struggle between and among opposing forces.

Running the gauntlet was a practice of some American Indian tribes and many military organizations. It was sometimes an initiation but more often a punishment for breaking a rule. The victim would use whatever speed and dodging ability he had to make his way between two parallel lines of men armed with clubs or rifle ramrods. He would try to reach the end of the painful pathway with as little injury as possible.

From planting through long weeks of cultivation and harvest, transportation and storage, and a multitude of uses, Indian corn was destined to run the gauntlet of an incredible number of enemies. To be sure, they were enemies of attraction rather than dislike: winged and crawling and climbing creatures that found the grain or the leaves, roots, or stalks irresistible. Not only was corn entirely at the mercy of human planters and tillers for the very survival of its species, but it also depended on people as its protectors against these hungry hordes.

And what a job of protection those corn farmers had. Predators from microscopic size to bear and two-thousand-pound buffalo worked around the clock and around the calendar to threaten the corn plant and grain. The multitude of its enemies seems to be a measure of corn's edibility. Observers in colonial times noted that rats, mice, and crows would leave other grains untouched if maize was to be had, and there were countless other creatures that had the same tastes. "Eternal vigilance" might as well have been written about the protection of the corn crop as about liberty.

Corn was most vulnerable when in the field, from planting to harvest. After it had been stored it was both sheltered and concentrated. Thus it was easiest to watch over at the very time when it required the least watching. In the patch it was much more easily reached by predators and too scattered to patrol and protect all at once. There was no better example of this than the settlers of the 1700s who could be seen firing their shotguns at huge flocks of blackbirds devouring their newly planted corn crops. The birds merely flew to the other ends of the fields and resumed their eating. When the gunners pursued and fired at them again, they flew back to where they had been eating before. And so on, until the poor farmers and their ammunition were exhausted.

The best answer to predators in the fields was the patrol, commonly made up of women, children, and the elderly by day and men by night. A large family meant many stomachs to feed, but it also provided a labor force and a big group to patrol the fields. Indian methods of patrolling, which had developed from long experience as had so many of their ways of coping with corn, were widely used by the pioneer settlers. The noisemakers, weapons, and superstitions were different, but otherwise the tactics were similar. Children, including young ones supervised by women and aged grandparents, were positioned at scattered points in the corn patch, beginning at daybreak when the field sparrow trilled reveille. This was particularly necessary at planting time and for several weeks after and later as ears began to grow on the stalks. These were the times when the numbers of daylight enemies were at their peak. Eulalia Pérez told historian Hubert Howe Bancroft that at the California missions children and pregnant women kept the birds away from the crops.

Young and old patrollers alike, occasionally from crudely built platforms in trees, shouted, waved their arms, threw rocks and clods of dirt, beat on pans, and whistled to frighten away birds and animals.

The ubiquitous scarecrow.

John Smith directed his Jamestown colonists to build a four-post platform in the center of each cornfield for this purpose. Sometimes the sentinels tied strings to many stalks and stretched them to a central spot like spokes of a wheel. With this device one person, by pulling the proper string, could shake a stalk or orchestrate a noise at the right location and thus scare pests away from a wide area. The settlers had the advantage of tin pans and occasionally bull fiddles as noisemakers. They also whittled whistles out of small hickory or ash limbs.

Settlers fresh from England called them shoy hoys. Immigrants from France called them images. To the old-time pioneers they were scarecrows. All birds feared humans, and the next best deterrent to having people in the cornfields was to make the birds believe that they were on hand. One answer was the manikins or scarecrows. These imitation people were very common, although their effectiveness was questionable. They were made around cross stakes, the long limbs driven into the ground. Here the earlier pioneer farmers were somewhat limited by scarcity of clothing. They had to save and wear even the tattered garments. Hence the widespread use of straw figures to frighten birds.

Also widely used but of doubtful effectiveness were objects suspended by strings from poles. Swinging shingles, painted white on one side and dark on the other, fluttered in the wind. The day patrol suspended dead birds at various points in the field. The glittering of sunlight from pewter plates twisting in the breeze was believed to remind birds of the reflecting gun barrels of previous battles. Other supposed reminders of the feared firearms were rags or old shoes filled with brimstone (sulphur) or soaked in oil and gunpowder mix, partially burned, and tied to poles in the fields. These smelled like gunpowder. They were most effective if posted on the upwind sides of corn patches.

And there were guns. The settlers fired at the winged and furred tormentors when lead and powder were not too scarce. Smoothbore muskets and, by the late 1700s, "rifle guns" were used against larger animals. Shooting a scatter-load of small shot was a favorite way of dealing with huge flocks of birds, for the farmers benefited triply. Their youngsters used the gunnery to improve their aim. "Most farmers have a son who would rather be shooting a gun off all the day, than be at plough or harrow," wrote a European visitor, and crop protection was of untold importance in developing accuracy with weapons. The booming noise helped to scare away pests. If the aim was good, meat was placed on the table since many of the creatures killed—including squirrels, rabbits, blackbirds, and sometimes crows—were eaten unless the family had an oversupply of meat at the time.

Indian children too used the corn patrol as a rehearsal to improve their shooting ability for later hunts. Traveler William Bartram reported that these youngsters, using bows and arrows, would "load themselves up with squirrels and birds" while protecting the maize. The Indians were more hampered by superstition than settlers because many of them believed that crows and blackbirds had been responsible for giving them corn and beans. To kill them would be to invite the wrath of the Great Spirit. In some tribal areas these birds were allowed to eat the crop unmolested. Roger Williams of colonial Rhode Island observed that not one Indian in a hundred would kill a crow. He might have added that not one crow in a hundred would pass up a corn feast.

Birds were the greatest threat at planting time, owing both to their habits and their unbelievably immense numbers. Since these aerial squadrons raided by day only and were not a direct threat to personal safety, the shooing and often the shooting were assigned to the physically less strong members of the watch force. Slaves presented a special problem, for although they had to be sent out to protect the crops, their use of guns was generally (but not universally) prohibited for fear of revolts. Men and older boys more often took over the night patrol. With the exception of birds of prey such as hawks, owls, eagles, and shrikes, most birds liked corn. Even feathered raiders as small as sparrows would fly in and peck holes in immature ears, causing them to spoil. Woodpeckers and waterfowl were particularly destructive at harvest time. Among the consumers of ripe corn were game birds, including passenger pigeons, Canadian geese, pheasants, wild turkeys, partridges, prairie chickens, quail, and doves.

Doves, whose range covered the entire country from coast to coast and border to border, are still called by some hunters the "gray ghosts of the corn fields." These birds were less noisy than most species, but were considered to be "the most cunning and mischievous of all." They liked both the tender spear and the moist grain at the bottom. One dove might "trip . . . from spear to spear until she has got twenty or thirty in her craw. . . . The crop is almost wholly destroyed. . . . If I must have one of the two, give me the cunning serpent in preference to the *harmless* dove." Fortunately for those on patrol duty, doves and passenger pigeons would usually fly into the fields twice daily, at fairly predictable times in morning and late afternoon. Thus guardians could slip away from the vigil occasionally, particularly from late morning to early or mid-afternoon when all raiders were less likely to be active.

Most damaging of corn's feathered enemies were crows and maize thieves. Settlers knew the latter by other names: jackdaws (because they looked like European birds of that name), blackbirds, and (officially) grackles. Crows and maize thieves were wise enough to pull and eat the tiny shoots of corn and also eat the soft moist kernels at the bottom thereby nipping the crop in its infancy. Cranes also attacked corn in this way, but they were less numerous except in fields close to

marshes and watercourses. Crows had some additional wisdom working for them. They would post guards on a rotating basis to warn the
eaters of approaching danger.

There were occasions when maize thieves came in such clouds that
they blackened the sky. At other times they flew in endless black
bands, stretching from horizon to horizon for several hours at a time,
hell-bent for some far field of sprouting corn or some thick grove of
cedars in which to roost. Colonial governments paid bounties for dead
crows, maize thieves, and woodpeckers as an added incentive for
farmers to kill them and protect the crops. The hunters would pull off
the heads of the birds as evidence to collect the bounties, saving the
breasts, legs, wings, and giblets of many for eating. They slaughtered
so many blackbirds in response to New England payments that this
species of birds in the region was nearly exterminated. Then in the
summer of 1749 worms destroyed the hay crop because there were
not enough birds to eat them! Devout Puritans were "conscience-
stricken because they believed this to be a punishment for meddling
in the providence of the Almighty Creator." Thus white beliefs in the
supernatural were involved along with those of Indians in the control
of these invaders from the air. Then, as in later times, humans dis-
rupted nature's balance when geese, cranes, squirrels, and maize
thieves knew better.

Settlers defended the seeds they planted by using an herb that induced a stupor in birds, by coating the seed with distasteful substances such as tar, and by planting the grains up to six inches deep.
Kernels poisoned with hitch root killed some birds. Destruction of
nests was another tactic. But nothing succeeded as well as guns.

Although most of the corn-eating mammals did their damage by
night, there were many that challenged the human patrol in broad
daylight, particularly in early morning and late evening. Squirrels and
chipmunks sometimes dug up newly planted grains, but were most
destructive when the crop matured. The gray and black tree squirrels
and their larger relatives, fox squirrels, often ate from ears on the
stalks, making such a racket that it was easy for a patroller to steal
down the rows until within gun range. These agile creatures with
flicking, bushy tails and sharp teeth would also gnaw off ears and
carry them to trees. There were piles of stripped cobs at the feet of
some trees near cornfields. Nailing or tying ears to trunks was a practice on many farms. The squirrels would congregate at these sites in
the trees and could be easily shot. A still-hunter might stand in one

spot for a short time and kill as many as a dozen squirrels at the same ear of corn before picking them up.

Sometimes squirrels descended on the corn crops in such numbers that companies of hunters would invade the neighboring woods and offer prizes to those who killed the most animals. These events were frolics involving a great deal of merriment. In recent years, with these animals no longer a threat, farmers have been known to nail corn to the trees as feeders. This saves corn in the field, since squirrels rip open many ears on the stalks, causing them to spoil. Colonies and, later, states put bounties on squirrels. According to Ben Franklin, the legislature of Pennsylvania colony had to cut the bounty from three to one and a half pence per animal in 1750. The £8000 for this expense alone had bankrupted the treasury.

Probably no meat in the entire frontier experience was more relished by settlers than squirrel flesh. Farmers even argued for using tails instead of heads as proof of kills for collecting bounties since the brains, jowls, and tongues were considered delicacies. Squirrel pie was a favorite dish at frontier work frolics.

Rabbit meat was rated almost equal to squirrel for the dining slabs of the frontier. Rabbits, moving out of brush and woods in early morning and late evening, hopping and stopping, hopping and looking, would nibble the tender corn plants to the ground. They would eat young shoots from first sprouting to foot-high growth. "One hare will nip off a whole row, in one night, forty or fifty rods long," commented English-American farmer William Cobbett. Many of these eastern cottontails were taken with the arrows and snares of the Indians, or the guns and traps of the settlers. In addition to steel traps which were universally used for fur-bearing animals on early American farms, a favorite device was the box trap, which was especially effective with rabbits. Unlike many of corn's enemies, rabbits increased in number with the clearing of the forests.

A heavy percentage of Americans today would find it unthinkable to shoot and eat the cute, busy little squirrels and rabbits and gentle-eyed does and fawns. But in early times, practically all the rural male population above nine or ten years of age and many of the women and girls were hunters and killers of wild things. For good or ill, it was a large part of the way of life. This was an age of a kind of "manifest destiny" attitude toward creatures, human and animal. All these citizens of the wild community were forced to give way before the American settlers regardless of who had first claimed the land.

"My enemies are worms, cool days, and most of all woodchucks," wrote Thoreau of his bean and corn gardening. Groundhogs or wood-chucks (American marmots) were very destructive in maize fields at the milky stage of the grains. These too were day raiders. They would stand on hind legs, bite the stalks, and break them over at the bite point to get at the ears. Usually they would eat enough to spoil the ears, then move on to other stalks and repeat the process, destroying much more than they used. Farm folk considered the meat of these burrowing animals edible, particularly that of the young chucks.

Across the Mississippi, prairie dogs and ground squirrels also took some toll of corn by nibbling off the young plants and digging up seeds. Hamlin Garland tells of young boys lying in wait with snares at ground-squirrel burrows. Whenever a head poked out, they jerked the line and another corn thief surrendered to the choking noose. The squirrels (erroneously called gophers) even learned to dig for corn seed wherever planter marks crossed. They were such a threat that the settlers had to "wage remorseless war upon them from the time the corn was planted until it had grown too big to be uprooted." Their weapons were shotgun, snare, and poison. In later years, poisons would wipe out millions of rodents and disrupt the balance of nature, driving black-footed ferrets to the point of extinction for lack of their natural food.

Domesticated ferrets, house cats, poisons, and a host of other rem-edies were not enough to prevent mice and rats from becoming the most destructive of all of the mammals that raided corn crops over the centuries. One rat might eat more than fifty pounds of corn in a year. A number of varieties of mice and rats (some of which had come as stowaways on colonial voyages from Europe) worked day and night on the hard, dried grains in shocks and in cribs and barns. There were structural ways of coping with them in the storage buildings with measures such as inverted pie tins. Some settlers employed ferrets to catch rats in their own ricks and cribs as well as those of their neigh-bors. Locking cats in barns at night was a widespread practice. Holes gnawed by mice were often plugged with corncobs, as were knotholes in boards. A nineteenth-century farm almanac advised farmers to heat plaster of paris in an iron vessel, and mix half-and-half with Indian cornmeal. After it was eaten it would "set" in the stomachs of the var-mints and kill them. Mouse and rat traps of several designs came on the market. Settlers did their cause a big disservice by declaring war on "chicken" hawks and owls, which were worth many times their weight in the finest treasure as enemies of rodents. Because of their great numbers, small size, and sharp teeth, mice and rats probably destroyed nearly as much corn as all other wild animals combined, particularly in the shocks.

One of the most destructive of the night prowlers was the raccoon, or "'coon" as it was universally called. Like the woodchuck, it preferred milky corn, although it would eat the mature grain in season. Dogs in the field at night provided the best prevention. A 'coon was a plucky fighter but preferred to flee the yapping pursuers and take to a tree. This was often a fatal mistake since it could be shot and used to fatten the pioneer's larder. The fur, however, was not prime at that time of year. Furs were taken from late October to March, when the raccoons had grown thick, warm undercoats.

As if there were not enough small creatures to menace the maize crop from the air and on the surface of the ground, there were several mammals that attacked from below. Pocket gophers in the South and the Mississippi valley ate the roots and plants of corn, pulling small plants down into their burrows. Most of the corn country was inhabited by the eastern moles, and although they ate chiefly worms and bugs, they also fed on softened seed kernels. Their burrows caused erosion of much corn land and disruption of crop roots since the root area was where worms were hunted by moles. One bit of strategy used on moles was to open up air holes in tunnels with hoes, then have a child stand guard at each until the little burrowers came to stop the air draft. It was then easy to pull them out of the ground with a stroke of the hoe. Other control methods included trapping and placing poison-soaked seeds and concentrated lye in the burrows. Deep cultivation destroyed mole burrows, and stumps were pulled to deny the tunneling animals a favorite refuge.

Nature has an uncanny way of balancing its accounts. The small creatures were on hand in countless millions for their own corn harvests. True, they had big appetites for their weight, but had they the stomach capacity of deer, bear, elk, and buffalo, corn never would have survived. Fortunately for the crop, the natural balance also worked the other way. Those animals with large appetites were only a fraction as numerous as the small ones, although it did not always look that way on the buffalo plains. Farmers shooing blackbirds from corn crops might have thanked their lucky stars that deer could not fly.

Among the large wild animals that threatened the crop of corn, the Virginia or white-tailed deer was the most widespread, covering nearly the entire corn country and eating both the leaves and grain of plants. Beginning while it was still daylight in late evening, through the night, and into morning hours, deer frequented the fields from first spring sprout until fall and winter husking and storing. They were not an unmixed evil since pioneers placed a high premium on venison and buckskin. As in the case of many other animals, the attractiveness of corn to deer has been overlooked by historians. Like the salt lick and watering hole, it was a major factor in hunting. The deer came to the hunter, who could shoot from hiding behind specially made cornstalk blinds during low-sun hours. So many of these animals were killed (not always in cornfields), particularly for the commercial hide trade, that the first colonial game laws other than bounties were for the protection of the white-tailed deer. Fortunately the legal shelter came in time for the white tail.

Except for the threat of white and Indian attacks on each other, physical danger in the corn-producing area, even at night, has been exaggerated. Despite all the fables about attacks by panthers (painters), wolves, and black bears, the only wild animals that had to be generally feared were grizzly bears. Their range was such that corn-growing settlers had almost no contact with them except in the Far West, and even there, contact was minimal. The others, including panthers and black bears, attacked humans only when wounded, cornered, or protecting their young. Indians of the Far West, having both grizzly bears and an occasional buffalo bull to face, found greater danger in protecting their corn. We have seen previously that the Indian range for corn growing was limited by buffalo on the plains.

Before 1800 buffalo posed some threat to corn throughout their range from the Appalachians westward, as did elk. By 1820 all of the buffalo and most of the elk east of the Mississippi had been killed. The inquisitive pronghorn antelope liked to graze on corn in the trans-Mississippi country, but was not a serious menace. In the Far Southwest the sometimes fierce collared peccary, a species of wild hog, did considerable damage to the Indians' corn crops, and in the tropics of Central America, monkeys were pesky maize eaters.

Pioneers depended heavily on their dogs to help scare off bear and deer from the grain fields and gardens. Many a settler camped in the family cornfield at night, accompanied by dogs and gun. The baying of dogs and firing a shot or two would usually send a bear scurrying. Black bears were night raiders, eating corn ears chiefly in the tender stage. Dogs were also useful in the ceaseless battle to keep domestic livestock from destroying corn crops. But an experienced northern farmer echoed the sentiments of most successful corn growers when he wrote in 1822, "I had rather have one length of good fence to keep cattle out of mischief, than all the dogs in the universe."

Frontier fences were not very effective in keeping wild animals out of cornfields. Deer bounded over and bear climbed across. That old frontier expression, "pig-tight, horse-high, and bull-strong" applied to very few fences before the late 1800s. Wooden rail fences, known variously as Virginia, zig-zag, or worm fences were somewhat useful in warding off domestic stock, especially when topped by a stake-and-rider structure. Post-and-rail fences took up less land and did not require as much wood, since they were built in a straight line instead of zig-zagged. Fences of rock were popular in the glaciated country of the Northeast, and on the prairies where wood was scarce, sod was tried. It met with as little success as the moat-and-castle technique of ditches and dykes. Some settlers made fences of lines of jagged stumps uprooted from clearings. Others fastened rails to tree stumps cut high above ground in the clearing process. A prominent Missouri farmer in the 1840s invented a portable wood fence which had some utility.

Building a fence was one thing. Maintaining it was quite another. Rot and insect damage to buried sections of posts caused the biggest expense of upkeep. Fence builders countered the problem by coating post bottoms with preservatives like turpentine or creosote, a derivative of wood tar; by charring the lower parts of posts; and by using the most durable woods, such as oak, red cedar, and bois d'arc or Osage orange.

That thorny tree of the deep yellow heartwood, Osage orange, was uniquely suited to fencing of a different sort. Professor Jonathan D. Turner of Jacksonville, Illinois, was the greatest American promoter of English-style hedgerow fences. He adapted Osage orange trees to prairie conditions where fence wood was scarce as other innovators used rose hedges in Texas. Although they suffered when weather was unduly hot or dry, these barbed hedge rows were by no means rare. One could tell how far the plains frontier extended before the invention of barbed wire by the western edge of Osage hedges.

From barbed branches to barbed wire was but a step. Ultimately, the most practical barriers were the various types of wire fencing. Plain and woven wire fences were in use on the treeless prairies by the 1850s, but they were brittle and rust-prone and often broke. Improved strands of wire arrived so late on the scene that their use was confined mainly to keeping domestic animals in check. Barbed wire was introduced in the mid-1870s and patented at about the same time by two inventors of De Kalb County, Illinois, Joseph Glidden and his neighbor, Jacob Haish. It caught on quickly and soon became a major factor in the range feuds between cattlemen, who wanted to let their stock roam freely, and sodbusting "nesters." Hundreds of inventors secured patents for barbed wire, but only a few designs were successful.

In New Mexico, where corn patches overseen by rimrocks contrasted with those overshadowed by dense timber far to the east, fencing presented different problems. Building fences was more difficult for the same reason that clearing the land had been easier—lack of wood. Josiah Gregg noted that the *labores* and milpas (cultivated fields) of New Mexico were seldom fenced. Such fences as existed were "poles scattered along on forks, or a loose hedge of brush. Mudfences or walls of very large *adobes* are also occasionally to be met with." Hacendados and rancheros in Spanish and Mexican California had their Indian laborers build fences of various types—moats or ditches, stone walls, and as in New Mexico, adobe fences. Some of them embedded cattle skulls on top of the adobe, leaving the sharp horns extended outward to discourage deer, bear, and other animals from going over. "Human fences" of Indian employees were assigned to frighten away birds and animals.

The fences that were too scarce, too low, or too dilapidated to keep wild animals out of cornfields were just as ineffective against domestic livestock. Open range was the rule in newly settled areas and along roadways in older, established farming country. When wild animals entered cornfields there was no one to blame, but livestock were owned by people, and often the owners could be identified by brands or ear clips. Let these animals wreak havoc in a field or garden, and there were grounds for a feud. Woods and prairie disputes of this type long preceded the notorious range wars of the Great Plains farther west. In fact, problems created by loose stock and unfenced fields were the greatest causes of quarrels and legal tangles between farmers, "whiskey not excepted." And since most Americans were farmers, these were probably the most common legal problems in the history of the country before the automotive age. Robert Frost was right: "Good fences make good neighbors." Liability for stock damage to crops varied with place, time, and situation. Generally, in the common-law domain of Anglo-America farmers were more responsible for protecting their own crops, while, as Gregg observed, New Mexico herders were responsible for controlling their livestock.

Among the countless thousands of pieces of advice published for farmers in those tillers' bibles, the farm almanacs, none were more pointed and oft-repeated than precautions about fencing. "Fences, fences, see to your fences," was a typical example. "If these are neglected, you may as well neglect ploughing, planting, and every other requirement on your farm." And there was much moralizing in parable form: "While Ichabod slept, the cattle of the stranger were devouring his cornfields." Even at the nursery-rhyme stage, "Little Boy Blue" was reminded that the sheep was in the meadow, and the cow in the corn.

Good fencing by all parties would have done much to keep the peace, but fences were expensive, and it was often necessary to shepherd the stock with both human and canine herders or drive the animals away from the fields by use of patrols. As did deer, elk, and buffalo, the settlers' horses, mules, donkeys, cattle, and sheep liked the taste of corn at any stage and had to be controlled. The domestic hog presented a problem of a different type, for it was not always domestic. It was usually allowed to roam freely, and it reverted very quickly to the wild state. Often at butchering time in November, it was hunted and shot like a wild animal. The hog's ability to root under fences with its nose was annoying. Hog farmers would clip wire rings in the upper edge of each porker's nose or would cut several sinews in the animal's nose while it was young, making its nose too tender to root.

Not that hogs were limited to getting through fences with their tough snouts. Richard Parkinson, who toured America at the end of the eighteenth century, described how these pesky beasts demolished rail fences. "They will go to a distance from the fence, take a run, and leap through the rails three or four feet from the ground, turning themselves sidewise." Once the fence was broken, other stock would go through to devour their favorite feed, corn.

While much corn was intended for domestic livestock, farmers planned to feed the crop under controlled conditions, preserving a large part of the grain and fodder until late fall, winter, and early spring when other forage was scarce. If hogging down was to be the method of harvest, at least the settlers wanted to plan it that way—and with their own hogs.

There was clearly a danger to personal safety from domestic livestock. Sows with young pigs and boars at any time were fearsome creatures, combining courage with deadly tusks. Bulls, too, were often quick to charge. Many more settlers were killed by the tusks of hogs, the horns of bulls, and the flying heels of horses and mules than by all the wild animals of North America combined. These domestic animals were more dangerous in rank corn growth than in pens.

Farm animals were easy to detect in the cornfields. Their breaking of stalks and crunching of grain could be heard from a distance, and their large size made them easy to see by crouching low and looking down the long straight rows. This was one disadvantage of contour farming; the rows were crooked.

These have been the enemies of corn that, for the most part, are of the passing and past scene. Birds are scarcely a continuing threat, although the blackbird and a successful twentieth-century immigrant, the European starling, may make another run at corn plantings. If so, they will be quickly destroyed by chemical warfare, which has already proved effective. Mice and rats remain abundant, but the banishing of shocks by combine harvesters and the building of tooth-proof storage bins have sharply cut their inroads on commercially grown corn. With the advent of modern wire and electric fences, livestock have been screened out. And to an almost tragic degree, many of those species of wildlife that were once serious contenders for the fruits of the corn harvest, like the early tools of corn culture, are confined to quarters where human onlookers can file by and view them or their stuffed remains as curious remnants of a day long gone. Worthy foes they were, although hated at the time, and they will be missed.

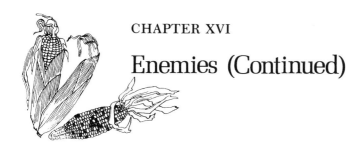

CHAPTER XVI

Enemies (Continued)

The contrast (with respect to diseases) with wheat and
most other grains is so strikingly in favor of corn as to
justify the conclusion that the exemption of the latter is
purposely ordered by a beneficent Providence.

EDWARD ENFIELD, 1866

Looking down the long straight rows, listening for
bird calls or the tearing of husks and trampling of stalks by animals,
turning dogs into the fields, or firing guns may have been useful tactics in coping with many enemies of corn, but a wide range of destructive agents could not be stopped by these weapons. They included
insects, fungi, weeds, foul weather, and marauding human beings
themselves. Unlike birds, wild animals, and livestock, all of these are
either serious threats to the corn crop today or are poised and ready to
become so with a change in conditions. And despite the fact that corn
was a more predictably successful crop than any other grain in the pioneer era, insects and diseases annually claimed an annoying percentage of the yield, often enough to spell the difference between profit
and loss for the producer.

Among the more than seventy insect enemies of corn, five were
most often mentioned by farmers of pioneer America: the armyworm,
cutworm, chinch bug, grasshopper, and weevil. Identifications were
not always accurate. Armyworms, for example, were among the thousands of subspecies of cutworms, and the worst so-called grasshopper
plagues were actually invasions of locusts. Armyworms were insects
in the larval stage. Sometimes they were so numerous that they
seemed to advance like an army, devouring every plant in sight. The
very name *armyworm* struck fear into the settlers' hearts, for there
was usually some talkative neighbor who claimed to have seen them
wipe out whole fields of crops.

Piling a row of corn fodder or straw ahead of the line of worms and setting it afire was one method of attempted armyworm control. Farmers improved on the burning tactic in the post–Civil-War era by use of "coal oil" (kerosene). At about the same time, they began baiting worms with poisoned bran. They put the torch to weeds along stone or wire fences in an effort to control these and other insect species. The settlers had to be careful not to damage fence posts, particularly the highly flammable but rot-resistant red cedar. Colonial farmers used dressings of sea manures, including salt water, to sprinkle around plants as worm repellents. Other preparations applied in this way included refuse brine from salt meat, water-and-walnut-leaf steepings, lime, soot, ashes, and finely chopped tobacco. Deep furrows, tar trails, and later creosote trails around corn fields discouraged worms. Some farmers would dig shallow post holes on the outside edges of these trails. Worms that followed along the trails looking for crossing points would fall into the holes and die in the heat of the next day's sun. By the time of the Civil War, inventors had devised wheels with sharp spurs for worm control. Worms followed along the smooth wheel tracks and fell into the spur holes where they died in the heat of the day.

A widely used early method of controlling some types of cutworms was to look for small corn plants that had been cut off, then search under clods or dig around the plants' roots until the worms were found. Another control tactic involved planting a small piece of potato next to the corn grains and examining it each morning. Worms preferred moist potato to dry grains. The workers would kill each worm and replant the potato sections to trap other worms. These methods of control were tedious to be sure, but by such labor frontier life was sustained. It must be remembered that this was not farming of giant fields to feed great city populations, but single-family survival farming.

In 1785 the chinch bug came to North Carolina, probably on a ship from a Latin American port. This was its first known appearance in the United States. It soon extended through the grain belts of the country, doing great damage by sucking juice from stalks of corn and other grass plants. The chinch bug was an adaptable creature, switching from dry wheat stems to green cornstalks in midsummer. Growers tried various methods of control including burning grass, weeds, or other cover where the bugs sought refuge, plowing furrows around fields, surrounding fields with boards set on edge and tarred on the top edge, and laying tar-soaked twine around the perimeters of fields.

Holes along the furrows, boards, and lines trapped some of the insects before they reached the leafy crops.

Grasshoppers and their look-alike relatives, locusts, infested pioneer corn crops, costing farmers untold millions of dollars. Dry summers seemed to be the worst for invasions of these winged, hungry hordes. The season of 1789 was a severe one in New England for locusts. They appeared in "endless swarms." When leaves and grain were consumed, they ate the bark of trees and shrubs. John R. Cummins of Eden Prairie, Minnesota, must have felt that there were winged serpents in his Eden when he wrote in 1857 about his experience with grasshoppers. "In the corn fields they would eat off stocks 8 or 10 feet high, and after it fell clean it all up."

The most memorable siege occurred on the Great Plains in the summers of 1874, 1875, and 1876, when swarms of locusts caused some $200 million damage to agriculture. Clouds of these creatures blackened the skies, devoured entire crops, and caused railroad locomotives to spin their wheels helplessly in the oil of the insects' bodies on the tracks. Farmer Annie Bingham on the Kansas frontier reported that the settlers, knowing their corn crops were doomed unless they acted quickly, cut the green stalks and saved the fodder even though ears had not yet developed. Those plants that were not cut were stripped down to bare stalks by the invaders. Farmers used a bait of bran mixed with Paris green, but a more important locust-control measure was increasing turning of the soil. Plowing for both corn and spring wheat early in the year and for winter wheat in the fall buried insect egg deposits too deeply for hatching. Birds helped to keep grasshoppers and other insects in check, but as in the case of maize thieves in New England, people were inclined to upset the balance of nature. Where hoppers were not too numerous, they could be killed by farm children who swatted them with light wooden paddles. Cornfields were favorite places to catch grasshoppers for fish bait. Locust and grasshopper plagues on the plains in 1936, 1979, and 1980 remind the country of its former struggles.

Others among the formidable insects with which settlers had to reckon were the grain weevil and its relatives, the rice or black weevil and the Angoumois grain moth. Their eggs, deposited in or on kernels of corn, hatched into tiny larvae, which bored inside the grains either in the fields or later in storage cribs and bins. The airtight log and clay-chinked cribs designed to control these little borers have been previously described. In the late pioneer period, fumigation of the sealed bins with chemicals such as carbon disulphide proved effective.

Numerous other insects plagued the early corn farmers. Wireworms (larvae of click beetles) and white and brown grubs ate germinating seeds and the roots of maize during the cool of springtime. Corn-root lice did extensive damage. Rootworms and budworms attacked the corn plants when the weather was warm. Mealworms consumed stored grain. There were stalkworms and earworms of several kinds, and corn flea beetles, which caused wilt disease by nibbling holes in leaves. In the Gulf states, sugarcane borers varied their diet with corn. Some farm families went so far as to examine each tassel, ear, and stalk for borer droppings in order to cut down the damage and kill the worms before they spawned new generations. Slugs were fought with trails of lye.

It was not unusual to find anthills in corn hills. The ants ate the moist seed, took moisture from the small plants, and "herded" corn-root lice or aphids, milking their sweet secretions. Gently stirring the anthills clogged their tunnels.

In 1917 an outbreak of the European corn borer occurred in New England. This worm, which probably had reached the United States several years earlier, spread west to the Rocky Mountains within a few years. Eating its way into stalks and ears of green corn, the borer, for nearly half a century, has done more damage to corn crops in the United States than any other insect.

Early American farmers at times thought they were fighting a losing war against the hordes of winged and crawling insects. They fought back with fire agriculture, which killed many bugs, but at the expense of needed soil fertility. They built bonfires at night to kill insects that were attracted and flew too near the flame. Some of the most effective weapons were deep plowing, which destroyed eggs, larvae, and pupae, and fall and spring crop rotation and interruption of cropping (that is, letting fields lie fallow for a year or two), which caused many insects to die by depriving them of their favorite meal for one or more seasons. One tactic, reported as successful by the *Genessee Farmer* of 1848, was to plant rows, and after worms had committed themselves, replant new rows between the earlier plantings. The worms were then largely killed by tilling out the decoy rows that had been planted first. Historians have neglected to give due credit to insects for their role in westward expansion. Bugs forced settlers to look for new squatter lands farther out on the frontier, where, for the first several years, the tillers of the earth could hope to keep a jump ahead of the tiny pests. Such moves interrupted plantings on many old clearings and thus had a cleansing effect.

Like insects, corn diseases came in many varieties. There were rots of kernel, ear, and stalk; and there were blights, rusts, wilts, and smuts. The worst of the fungus diseases were ear rot and smut. Of all the diseases, smut was the most obvious to the observer. It took the form of a large, silvery-coated pod or gall, which could occur anywhere on the plant and usually spread to the ear. In its mature stage the tumor was filled with a black dust. Many Indian tribes ate smut in its unripened, whitish stage, much as they ate mushrooms. Settlers partially thwarted this disease by removing and burning the affected plants as soon as the growths were discovered. Feeding to livestock resulted in the spreading of the spores through manure. Smut could be partially controlled by applying salt to the diseased areas of the plants. Overall, however, pioneer farmers had little defense against corn diseases except measures similar to those used to control insects—crop rotation, fire, removal of affected plants, fallow fields, and moving on to virgin land. Migration was the most effective, since diseases followed the advancing frontier more slowly than other enemies.

Modern corn farmers have a whole arsenal of weapons to bring to bear against insects and diseases. The use of poison bran and creosote paths around the fields in the late nineteenth century ushered in the era of scientific controls. Development of resistant plants has been of great importance, although hampered by the extinction of some subspecies through constant selection for highest yield. The careful timing of plowings and plantings and the application of potent chemical pesticide sprays such as aldrin and DDT have also been useful. Even plant inoculation is used. Yet the current loss to these scourges in the corn belt is about ten percent of each crop. Some of the control methods, chemicals in particular, have caused a new round of problems, such as poisoning of streams and groundwater. Some scientists believe that long-continued attempts at control by sprays breed races of superbugs that will be resistant to all presticides. Because they grow to adulthood quickly and reproduce many generations in very short time spans, insects can transmit survival characteristics quickly.

Weeds were another cornfield pest, but at least they could always be seen, and they did not fly or run away when approached. Farmers had hundreds of kinds of weeds to contend with, although in any given locality four or five were usually most troublesome. Grasses such as crabgrass were very short but matted in their growth, while horse-weed grew almost as tall as the corn itself. There were climbing pests, including morning glories and twining milkweeds. Other common weeds of the corn country were smartweed, goldenrod, fennel, cock-lebur, ragweed, lamb's-quarter, pigweed, sorrel, piper grass, bull net-tle, and jimsonweed.

For today's corn farmer, control of these weeds is almost ridiculously easy. Herbicides are sprayed on the soil at planting time, and perhaps again in a few weeks, and the chemical war is over for the season. Not so for the pioneer farmer. As described earlier, the means of control under normal conditions were simple but never easy. They demanded back-breaking work—hoeing each row; chopping the pesky plants; alternately cultivating lengthways and crossways along the neatly planted lines; mulching or loosening the soil to hinder future weed growth. Conditions, however, were not always normal. An untimely illness or an injury to a worker, whether human or draft animal, could put the family corn farm "in the weeds." Settlers' feelings ran strong against weeds, as they had against woodchucks, maize thieves, and the rest of the animate raiders. But there was another view, if only rarely held. Plow, hoe, cultivator, scythe, and mowing machine were partisan yet indiscriminate, cutting roots, nipping buds, and driving entire species of flora to extinction. A few who watched, though chop-pers themselves from time to time, dropped an occasional word of re-morse for the swiftness and heedlessness of the slaughter. John James Audubon, George Catlin, Washington Irving, Francis Parkman, Hor-ace Greeley, Peter H. Burnett, and Henry David Thoreau were among them. Farmer Thoreau had mixed feelings. "Have at him, chop him up, turn his roots upward to the sun," he said of the seemingly relent-less weed; then, in a more pensive mood, "But what right had I to oust johnswort and the rest, and break up their ancient herb garden?"

Then there was another villain among the variables, the weather. Thoreau referred to weeds as "those Trojans who had sun and rain and dews on their side." A prolonged rainy spell would gum the soil and make it impractical to hoe and impossible to cultivate with horse or mule. Once the weeds caught up to the corn, there was big trouble. Many were the families that went into the fields in these conditions and pulled weeds by hand to try, sometimes in vain, to check the growth without damaging the soil.

Nor could the skies be relied upon for kindly treatment after the corn was safely ahead of the weeds and "laid by." Weather was the "friendly enemy" of the corn crop. Its alternate sun and rain in moderation were as indispensable as its extremes were disastrous. The five hundred almanacs which flung out their words of contemporary wisdom to tillers of the soil in the early 1800s had as their most important single purpose the forecasting of weather. But practically speaking, they were of little help in preparing farmers for the capricious "nature of nature." Lincoln's earliest memory of life in Kentucky was heavy rain washing out the corn and pumpkin seeds he had just finished planting. There were floods in river bottomlands. Clouds rolled, rains fell, currents roiled, streams burst over their banks, and a year's labor was lost in one swirling wash of tawny water. A hot, dry spell, a threat at any time, was particularly damaging when it occurred at the tasseling stage. Silks would become brittle and perhaps fall off the ears, causing poor pollination. Curling or rolling leaves was the first danger sign. (American corn farmers, who appear to have lost about 16 percent of their 1980 crop to heat and drouth, need no reminders.) Likewise a high wind or very heavy downpour at tasseling time could hinder pollination. When plagued by such conditions, farmers would gloomily predict that their crops would be "nuthin' but nubbins." Winds also contributed to "lodging"—bad leaning and breaking of stalks. One June 27, 1821, Thomas Jefferson wrote of "a tremendous gail [sic] which "tore corn to pieces."

It was the unusual about the weather that pioneers feared, yet with 75- to 120- and even 150-day growing seasons, something unusual usually occurred. Late spring or early fall frosts cast a somber mood over the farm country. On small farms, beleaguered soil tillers combatted frost by building fires on the windward sides of fields. Deep and lasting snow blankets imposed double jeopardy on those who wrung their living from the soil—harder winter toil and a worse crop of summer insects because a heavy snow cover meant a light freeze underneath. Next to drouths, the farmers dreaded hailstorms most. Hail and gale usually arrived together. The pelting balls of ice could wipe out an entire year's crop in a few minutes, disrupting pollination and scarring green ears so that they later rotted. As often as prayers were uttered in the farmers' struggle against enemies, they were probably never more earnestly breathed than when the low clouds beneath a black thunderstorm took on the green-gray hue believed to foretell a hailstorm.

Men were known to create their own storms in the cornfields—the swinging of swords and sticks and lighting of fires among the crops of enemy peoples, and the hail of grapeshot and canister and musket balls. Many maize plots became battlegrounds. Scorched-earth methods of warfare are doubtless as old as human existence. They were widely used by both Europeans and Indians before the two met in the New World. Often it was safer, easier, and more effective to raid the enemies' crops than to meet in personal combat. Destruction of a year's corn and bean supply was a serious blow to the victims of grain robberies.

Warring Indians not only destroyed each other's standing crops. They thumped the ground with clubs in likely areas and listened for the hollow sounds that might reveal underground corn-storage hideaways. This trick was copied by settlers on the warpath against the aborigines.

The friendship of those initial meetings of the hemispheres at Jamestown and Plymouth soon waned. Indian tribes recognized that their maize lands and hunting grounds were being taken over. The marriage of Powhatan princess Pocahontas to Englishman John Rolfe at Jamestown seemed to assure good Indian-white relations, but when Chief Powhatan died in 1622, his successor Opechancanough led the tribe in a bloody two-year war to drive the invaders into the sea and safeguard the Indian lands. Groups of white settlers, protected from arrows and spears by armor, entered the Indians' strongholds each fall and burned stores of corn and beans after the harvest. The strategy was effective: twenty years of peace followed as the tribe was forced west to survive. Aged and infirm, Opechancanough again ordered an attack on the encroaching whites in 1644, but after much loss of life the settlers burned Indian grain, captured and killed the Powhatan chief, and forced the Indians to make peace and again move farther west.

In the Pequot War of 1636, both the Pequot tribe and its crops were exterminated by New England colonists. Forty years later an alliance of nearly all the New England tribes tried to roll back the ever expanding pioneers in King Philip's War. The opposing groups destroyed great quantities of each other's corn crops before the Indians made peace on the terms of the New England colonists.

This pattern was repeated again and again from the northern colonies to Georgia as the pioneers advanced. Corn crops were destroyed by each side to gain or keep possession of corn and hunting lands. It happened in the French and Indian War (1756–1763), and in 1774 the Shawnees under Chief Cornstalk bitterly resisted colonial "land stealers" in Kentucky and Ohio. Lord Dunmore's War of that year opened lands in these territories for settlers of the English colonies. Several years later during the Revolution, however, Shawnee attacks prevented corn planting by pioneers from Pittsburgh to Boonesborough, and the newcomers complained about their monotonous all-meat diet.

In the Chickamauga area to the South, Colonel Evan Shelby destroyed twenty thousand bushels of Cherokee corn in 1779. The Indians agreed to stop hostilities. Jefferson lost about eighty barrels of corn in 1781 "for want of cultivation by loss of the hands," slaves who left to join the British in the Revolutionary War. At about the same time, the soldiers of British General Lord Cornwallis ravaged parts of Virginia, destroying fields of corn and burning barns of the grain after they had taken all they could use. Far to the west, Yuma Indians of the Colorado River lowland killed many Spaniards in 1781 and permanently closed the vital Anza Trail from Mexico to California largely because of destruction of maize and other Indian crops by Spanish cattle and horses. Meanwhile, American success in the recently ended Revolution accelerated encroachment upon Indian lands west of the Appalachians. "Mad Anthony" Wayne destroyed miles of Indian maize fields in Ohio in 1794 and forced acceptance of the Treaty of Greenville, securing more lands for settlers. Corn raiding was likewise a major part of the strategy in the western theaters of the War of 1812.

The weapons of these raids varied with the nature of the assignments. Swords, ramrods, clubs, and horse-drawn log drags destroyed green corn in the fields. Corn shocks of pioneer farmers were veritable torches waiting for a flame; stored grain was surrounded with wood and burned or hauled to streams and dumped. Mutual fears between Indians and settlers made protection of the crops difficult. Ambushing the opponents as they worked in their own cornfields was not uncommon.

A number of battles of the Civil War were fought partly in rank growths of corn, including the bloody engagement at Antietam, Maryland, and the opening fight of the war in the Far West, Mesilla, New Mexico. Philip Sheridan's army so thoroughly plundered the Shenandoah Valley, burning corn and other crops, that it was said a crow flying from one end of the valley to the other would not have found enough to eat. And when William Tecumseh Sherman laid waste a pathway through Georgia and South Carolina, he gave careful attention to the southern staple, corn. Shocks laboriously stacked by women, children, and aged men were systematically put to the torch by

Sherman's "bummers." On the other side of the ledger, the usually underfed southern soldiers raided cribs and smokehouses in their own section during the great conflict. Confederates captured the Union steamer *J. R. Williams*, which was hauling 150 barrels of hominy up the Arkansas River for delivery to tribes in Indian Territory.

Corn thefts where people were heavily dependent upon the grain for survival were considered serious offenses. It was difficult to steal much from a shock, but stored grain was an easier target. The Aztecs imposed death by hanging as punishment for stealing maize unless the thief was poor, in which case there was no penalty at all. According to writers of farm almanacs, thievery was common. "Scoundrels, thieves . . . night villains," the culprits were called; the types "that more or less infect every neighborhood." Rather than have prescribed procedures for handling every kind of crop theft, it was customary to use the "watermelon law." In a statement reflective of widespread hostility against Latter Day Saints in western Missouri, Josiah Gregg wrote that the Mormons were "at first kindly received," but that the "good people" of the country "soon began to find that the corn in their cribs was sinking like snow before the sun's rays." Whiskey thieves in all areas learned to bore through floors and into barrels from below, catching the delectable nectar in waiting buckets.

Over the millennia Indian corn became the central issue in numerous wars of survival. Before the seventeenth and eighteenth centuries the seemingly unending forest, with its nuts, fruits, seeds, leaves, wood, and protective foliage, had enabled uncounted billions of birds, mammals, insects, and other creatures to live. Then came the clearing of vast areas of timber by settlers who had their own concepts and needs of survival. Corn was central to these. The removal of the forests drove the wild creatures to the brink of starvation, and for many, corn was a delicious alternative to the woodsy foods that they were being deprived of. Nothing short of life or death was at stake on both sides.

Thus far, people, with superior reasoning power and weapons, have won the struggle against birds and mammals and have reduced some species to extinction. Weeds and wars are currently (and somewhat uneasily) held in check by herbicides and peace agreements respectively. The smaller antagonists, such as insects, fungi, and bacteria, have not yet conceded that humans and their domestic livestock will win the great battle for food in the grain fields. Perhaps the worst threat of all has been humankind's well-intentioned development of high-yield corn varieties. These have swept the country and caused the neglect and disappearance of countless varieties that may be critically needed for their genetic diversity. The future of the world's food supply may hang precariously on a few scattered and primitive Indian corn plots with their disease resistant plants or upon the recent chance discovery of a variety of teosinte grass. In 1969 a hurricane and other unusual weather scattered a mutant strain of southern corn leaf blight throughout the Midwest. It destroyed 20 percent of the 1970 corn crop. Said Walton Galinat, research professor at the University of Massachusetts, "If we hadn't had surplus corn stored from previous years, many people could have starved worldwide. . . . The genetic variability was still available to give control by breeding for resistance. Next time we may not be so fortunate."

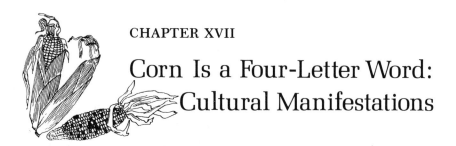

CHAPTER XVII

Corn Is a Four-Letter Word: Cultural Manifestations

As corny as a chorus of "Hearts and Flowers."

BILLY ROSE, 1957

I've been through the mill, ground and bolted, and come out a *regular down-east johnny-cake*, good when it's hot, but when it's cold, sour and indigestible.

Captain's statement to Richard Henry Dana and the crew of the brig *Pilgrim*, 1834.

If it is sentimental, old fashioned, unsophisticated, hillbillyish, or overly dramatic, it is *corn* or *corny*. If it is inferior bootleg whiskey, it is *corn, corn juice*, or *corn mule*. A drunk is *corn pickled, corned*, or *corn-fed*. One who holds to outdated styles or preferences is a *cornball*. When two trains collide head-on, they are in a *cornfield meet*. A country dance is a *hoedown*. *Corn on the cob* may be a reference to a harmonica, *corn shucks* to chewing tobacco, and *corn* to paper money. A stout person is *corn-fed*, and something good or highly useful may be labeled a *corn cracker*. *Corn muffin* is a restaurant or lunch counter term used in relaying an order. *Corn Willie* (corned beef or corned beef hash), *corn stealer* (a human hand), and *corn punk* (a variation on the simple term *punk*) are occasional colloquial usages. One who admits an error is said to *own the corn*.

Youngsters in pioneer times were called *nubbins* after the small, imperfectly developed ears of corn, and *tasseling out* was a term applied to a young person nearing maturity. A tough job was a *hard row to hoe*, and *rougher than a cob* described jagged surfaces. An emotional burden was a *millstone around the neck*. Some fact or idea that could be seized upon and perhaps distorted was *grist for the mill*. Gunpowder was *corned* if it was grainy, and a kernellike lump on a toe was a *corn*. To the southerner, *all hominy and no ham* was a way of saying *all work and no play.* In Civil War times, southern soldiers were so closely identified with corn diets that they were sometimes called *Corn-federates.* A *shuck* is a no-good person or object. And as an exclamation of disgust or disdain, or just an exclamation, *shucks* was one of the most common terms in the country's vocabulary.

Despite the fact that wheat is the number one crop in Kansas by a wide margin, no one would think of composing a song with a line such as "I'm as 'wheaty' as Kansas in August." The maize plant has loaned its adopted name to the cornflower (blue bottle), corn poppy, and the corn snake, a colorful reptile of the grain country. The nicknames and slogans of a number of corn-producing states reflect the preeminence of this crop. It was no accident that Indiana's Homer Capehart, son of a tenant farmer, and later a three-term United States Senator, chose a cornfield for his meeting to dramatize the 1938 protest against the New Deal policies of President Franklin Roosevelt, and called it a "Cornfield Conference."

Corn has found its way into the lexicon of American slang, with a range of usage unequaled in the annals of the nation's products, but such sounds of America talking are not the sum of the plant's cultural legacy. They are scarcely more than a suggestion. Any focus on the mix of corn and culture calls for a wide-angle lens, surveying humor, historical and descriptive literature, fiction, poetry, and music, as well as art, sculpture, architecture, and religion.

Perhaps psychologists have the most concise term for it—*Gestalt*, meaning an entire pattern of relationships. Corn as a socioeconomic and cultural phenomenon was a vast configuration, a unified whole to pioneer Americans. Often without their full realization, it evoked feelings of something good and necessary and pleasurable. Not only did corn-related thoughts conjure up pictures of play, work, profits, survival, and useful items around the farm, but they also touched and stimulated every sense-memory, often with a romantic hue. The sights of geometric green rows and the high-heaped golden harvest and the sounds of rustling leaves, of ears thumping rhythmically against bangboards, and of grains clattering into a tub below the sheller were music to pioneer spirits. The feel of dent-grained ears and yielding husks, the smell of steaming pudding and frying fritters, the taste of summer's first hot roasting ears dripping with butter, and not the least for some millions of Americans, the embrace of lovers secluded in a deep, forestlike hideaway of guardian stalks and blades—

these sensory responses wove themselves into a conscious and subconscious unity, a social panorama wherein, as in the meaning of the
harvest, the whole was indeed greater than the sum of its parts. Corn
was a complex of intriguing associations, a cultural catalyst, a cen-
tripetal force that drew the loose-knit, highly individualistic frontier
and agrarian society into a sense of common experience and common
purpose. In 1892, on the four-hundredth anniversary of Columbus'
great discovery, maize was adopted as the national flower of the
United States. So all-pervasive was its influence that its cultural legacy is alive and well today in an overwhelmingly urban scene.

In the process of composing this book, the illustrator and I have
often noted that the very suggestion of a volume on corn causes the
average person to chuckle and ask, "Are you serious?" Yes, corn is a
serious subject, but it is laden with humor. Perhaps the instant laugh
that its mere mention brings forth is a societal hand-me-down, a memory of corn-induced pleasures from the subconscious of America past.

Settlers in their cups at many a husking bee fell back on the old
frontier standby of bawdy humor. They sang coarse songs and, amid
pointed comments and guffaws, held corn ears or cobs aloft as phallic
symbols. As we have seen, a serious use of the corncob in early times
was for tubular toilet paper. If most Americans are unaware of this fact
today, they are familiar with the joke, for they have seen on the hall
wall of many a building the small red box with a white and a red corn
cob inside and visible through glass. The sign on the box is always the
same—"In case of emergency, break the glass!"

Perhaps it was only partly in a jesting vein that Dick Ranney, Iowa's
director of tourism, suggested using pigs and corn to attract tourists,
much as New York advertises its Great White Way and the far western
states their beaches and mountains. City people seldom see corn and
pigs, he noted. "Rows and rows of a well-plowed field can really be
something if you are from the big cities."

William James Lampton wrote of Kentucky, "where the corn is full
of kernels, and the Colonel's full of corn." Novelist Irving Bacheller
varied the story of "looking a gift horse in the mouth" with his account
of the type of person who, if given a bushel of corn, would say, "Won't
you please shell it for me." Jubilation T. Cornpone is a character from
Al Capp's comic strip, *Li'l Abner*, and Hank Ketcham's Dennis the
Menace says to an aggressive challenger, "I'll fight ya TOMORROW . . .
we're havin' corn-on-the-cob tonight, so I'm gonna need all my teeth."
In his "Corn Pone Opinions," Mark Twain included an opinion of his

own: "You tell me whar a man gits his corn pone, en I'll tell ya what his 'pinions is."

Probably no form of humor was more common in pioneer and rural America than the humor of exaggeration. Corn came in for its share. "Paul Bunyan's Corn Stalk" tells of the great fertility of North Dakota soils. The stalk grew so fast that Ole, who had climbed it to cut off the top, could not slide down because its growth rate was three times his speed of descent. Febold Feboldson, a legendary Swedish farmer of the Paul Bunyan mold, conquered unbelievably fierce blizzards on the Great Plains and weathered summers so scorching hot that corn popped right on the cob. A Cotesfield, Nebraska man picked up this theme during the 1930s, in his "Little White Lies" verses about Ben Gasst, a mythical teller of tall tales.

> So hot no more the hoe he'd wield,
> It popped the corn out in the field.
> We're getting close to hell, I fear,
> When it pops the corn upon the ear.

A New England almanac of the 1780s printed an epitaph of a miserly farmer who starved himself while hoarding corn. "Here Cornlay lies, in cold clay clad, who dy'd for want of what he had." John Langdon Heaton reported exaggerated stories of western corn families in his volume, *The Book of Lies* (1896).

> A good illustration of nature's bounty happened some time ago in Do-niphan County, Kansas. . . . A seven-year-old daughter of James Steele was sent, in the middle of the forenoon, to carry a jug of switchel to the men, who were at work near the middle of one of those vast Kansas cornfields. The corn was about up to little Annie's shoulders as she started, but as she went along it rose and rose before her eyes, shooting out of the soil under the magic influence of the sun and the abundant moisture. Almost crazed with fear, she hastened on, but before she could reach the men, the stalks were waving above her head. The men were threatened in a like manner, but by mounting a little fellow on a big man's shoulders, to act as a lookout, they managed to get out, when they promptly borrowed a dog, to follow little Annie's trail. It was not until late in the afternoon that they reached her, where she lay, having cried herself to sleep, with the tear-stains streaking her plump cheeks.

A mythical Dr. Binninger topped a boast of Jersey corn with the re-mark that it was

> nothing to what I have seen. In Gastley County, Missouri, I once saw the corn growing to such an unprecedented height, and the stalks so excep-tionally vigorous, that nearly every farmer stacked up, for winter fire-

wood, great heaps of cornstalks, cut up into cord-wood length by power saws run by the threshing engines. One man, Barney Gregory, took advantage of the season to win a fortune by preparing cornstalks for use as telegraph poles.

Another tall story from Missouri told of a corn shucker so swift that he could shuck the down row and two side rows in the cornfield without ever stopping the team and wagon. But one day he made a mistake. Grabbing at an ear from the down row, he missed and seized his own foot. Before he could stop his momentum, he tore off his shoe, threw himself into the wagon, and was hit by seven ears of corn that he shucked and tossed on his way in!

There were other gags of exaggeration built upon corn, such as, "If you store your corn in the ear, you must have awful big ears," and the question, "If a pirate is a buccaneer, what's a buccaneer?" The answer, "An awfully high price to pay for corn." Rudolph Erich Raspe, creator of Baron Munchausen, wrote, "Upon this island of cheese grows plenty of corn, the ears of which produce loaves of bread ready made." A definition of "corn licker" was given to the Distiller's Code Authority of the National Recovery Administration by Irvin S. Cobb: "It smells like gangrene starting in a mildewed silo, it tastes like the wrath to come, and when you absorb a deep swig of it you have all the sensations of having swallowed a lighted kerosene lamp. A sudden, violent jolt of it has been known to stop the victim's watch, snap his suspenders and crack his glass eye right across."

Samples from the literary passage on "Korn" give some idea of the style of late-nineteenth-century wag, Josh Billings.

> Korn is a serial, i am glad ov it.
> It got its name from Series, a primitiff woman,
> And in her day,
> the goddess ov oats and sich like.
> Korn iz sumtimes called *maize*, and it grows
> in sum parts
> of the western country, very amaizenly.
> . . . .
> I have knawed two hours miself on one side ov a korn dodger without produsing enny result, and i think i could starve to death twice before i could seduce a corn dodger. They git the name *dodger* from the immegiate necessity ov dodging, if one iz hove horizontally at yu in anger. It is far better tew be smote bi a 3 year old steer, than a korn dodger, that iz only three hours old. . . . Korn has got one thing that nobody else has got, and that iz a kob. This Kob runs thru the middle ov the korn, and iz as phull ov korn as Job waz ov biles. . . . In Konklushun, if ya want tew

git a sure crop ov korn, and a good price for the krop, feed about 4 quarts ov it to a shanghi rooster, then murder the rooster immejiately, and sell him for 17 cents a pound, krop and all.

Fortunately for the recording of history, not all early mention of corn was in the mode of humor. Corn has produced its own galaxy of literary figures for both fact and fiction. There are uncounted thousands of references to this great grain in the factual literature of pioneer America. Most are passing notations which are of very limited value individually but which, taken as a body, are indispensable in sketching the scene of corn culture and history. The more comprehensive treatments are few, and each is usually limited to one region. Several literary forms are clearly identifiable.

The farm almanac, that bible of agricultural crafts, makes up the most distinct genre. Corn lore took its place beside admonitions about weather, livestock, cookery, and other crops in the quaint pages of these rural preachings. There were hundreds of almanacs, the best example of which was probably the *Old Farmer's Almanack* which was published for several decades in the early nineteenth century. Since writers of these journals frequently repeated themselves and each other, agricultural lore (for it was not always knowledge) spread systematically and rather rapidly across the country. With some allowance for variations in region and latitude, the time-worn adage, "If you've seen one, you've seen 'em all" applied to the almanacs.

A second systematic type of reference, also involving much standardization of corn lore, was the farm journal. Here, too, there was much repetition from journal to journal. Writings from one were often summarized or quoted in others, and letters to editors from farmers were frequently quoted in the pages of journals as widely separated as Maryland, Maine, and Missouri. Among the informative letter writers were men from a class generally known as gentleman farmers or experimental farmers. The journals, in addition, provided excellent illustrations of early agricultural implements and machinery. The earliest prominent and continuing journal with national circulation was the *American Farmer*, published at Baltimore beginning in 1819. Other well-known journals included the *New England Farmer*, the *Genesee Farmer* (which merged with the *American Agriculturalist*), the *Ohio Farmer*, the *Wisconsin Farmer*, the *Prairie Farmer*, the *Southern Cultivator*, and the *Country Gentleman*, and there were many more. Since most of the nation's people were farmers, it is not surprising that agricultural journals made up a very high percentage of the periodicals. Nor is it surprising to find in them frequent references to the number one crop, corn.

For the same reasons that farm journals led in their field, newspapers stressed agricultural news heavily. It was a natural reflection of the predominantly rural population. Newspapers also copied wholesale from each others' columns. One agricultural columnist of the *New England Courant*, "Cornplanter" by pseudonym, was widely quoted in other papers and almanacs. A number of eighteenth century papers, such as *Farmers' Weekly Museum* (Walpole, New Hampshire), *South Carolina State Gazette* (Charleston), Pittsburgh *Gazette*, *Kentucky Gazette* (Lexington), *Centinel of the North West Territory* (Cincinnati), and *Mississippi Gazette* (Natchez), dispensed farm information by the 1780s and 1790s. Other organs picked up and continued the tradition in the early 1800s. From the *National Intelligencer* of Washington, D.C., to the *Missouri Intelligencer* at Franklin far up the river west of Saint Louis, from the *Niles Weekly Register* of Baltimore to the *Orleans Gazette* and New Orleans *Picayune*, from the *Indiana Gazette* of Vincennes to the *Missouri Gazette* of Saint Louis, the message was largely political and agricultural.

Encyclopedias and rural lexicons, although not numerous, carried needed advice to farmers. Information on such topics as Indian corn, fodder or stover, and the enemies of corn in Samuel Deane's the *New England Farmer* (1790) was useful and entertaining.

From time to time visitors from Europe or the eastern seaboard, and American planters themselves, wrote significant accounts of farming in general and corn in particular. Among the general treatments, Hector St. Jean de Crevecoeur's *Letters from an American Farmer* and Thomas Jefferson's *Farm Book* are outstanding sources. Frances Trollope, Frederick L. Olmsted, William Byrd, and Andrew Burnaby made observations on farm life in the West and South.

By far the most important of the early writings focusing on pioneer corn growing before the mid–nineteenth century were the accounts of two science-oriented European travelers to America. Pehr Kalm of Sweden visited the northeastern colonies in the 1740s and wrote the brief but weighty "Description of Maize, How It Is Planted and Cultivated in North America, Together with the Many Uses of this Crop Plant." Englishman William Cobbett, who at times wrote under the pseudonym "Peter Porcupine," was a voluminous producer of agricultural literature. His book, *A Treatise on Indian Corn* (1828), was written after he had spent a number of years as a corn farmer on Long

Island. It is the most extensive early account of corn culture, and although limited to a small geographical area, its information has much general value. P. A. Browne's "An Essay on Indian Corn" (1837) and spotty descriptions of life in the West from letters and memoirs of pioneer farmers carry the corn story to the mid–nineteenth century, after which the coverage of the subject became somewhat more complete and systematic.

Beginning in 1847 and continuing into the last half of the century, the United States Patent Office, Agricultural Section, issued annual reports of great importance to farmers. In 1862 the Department of Agriculture became a separate agency. The reports of the commissioner of agriculture (called the secretary of agriculture after 1889) continued to appear annually. This office dispensed seed corn as well as ideas to corn growers. After the Morrill Act of 1862, the various land-grant colleges of the states produced literature of interest to farmers. Prior to this time, almanacs, journals, newspapers, and agricultural commissioners' reports had served, albeit informally, as a kind of "mail-order university" of rural America. They were an important means of spreading crucially needed information, and they taught basic reading skills to the predominantly agrarian population as well. Though of mixed and often limited literary merit, these writings help to give a picture to analysts of the cultural and economic scene.

The half century following the Civil War produced a number of books on corn, with emphasis on "how to." Edward Enfield's *Treatise on Indian Corn* (1866) was the most comprehensive early guide to corn farmers. It was followed by William D. Emerson, *History and Incidents of Indian Corn* (1880), E. L. Sturtevant, *Indian Corn* (1880), H. Myrick, *The Book of Corn* (1903), E. C. Brooks, *The Story of Corn and Westward Expansion* (1913), and E. G. Montgomery, *The Corn Crops* (1915). In later years, books by Paul Mangelsdorf, Henry Wallace, and others have added a modern perspective. Aldo Leopold, a twentieth-century Thoreau, based his nostalgic requiem for the country's wildlife in cornfields as well as bogs, marshes and sand plains. "The drab sogginess of a March cornfield, saluted by one honker from the sky, is drab no more." In another passage Leopold observes, "Sand plains were meant to grow solitude, not corn."

It would have been surprising if American fiction writers had neglected such a popular American subject as corn. They have not. Countless authors of novels and short stories have turned to corn in some form and to some degree for background, foreground, or on occasion, central theme, of their settings. And why not? It was a common denominator with which readers in every state, every region, every section could identify. As in the case of factual literature, fictional references to corn are so voluminous as to defy listing and description. Little more than a representative sampling is practical in a work of this scope.

Beginnings were early but scattered. *The Adventures of Jonathan Corncob, Loyal American Refugee*, by Corncob himself (the pseudonym of an unidentified author, who was probably an English sailor in America), was a bawdy farce of the revolutionary era. It played more on the name than the theme of corn. A modern-day reviewer suggested an appropriate subtitle for *Jonathan Corncob*—"By Bundling Possessed." Harriet Beecher Stowe's moving novel of the cruelties of slavery, *Uncle Tom's Cabin* (1854), was cast in corn, cotton, and tobacco country. A generation later, Albion Tourgee, former carpetbagger, echoed the need for racial reform in his fiction on the Reconstruction era, *A Fool's Errand* (1879), and *Bricks without Straw* (1880). Inescapably, these accounts of rural North Carolina in the post–Civil War period touched upon corn culture, as did the writings of humorists Samuel Clemens and Joel Chandler Harris. Creator of the legendary spinner of tales, Uncle Remus, Harris referred to himself as a "cornfield writer."

Several fiction writers of the late nineteenth century focused more directly on the corn theme. Edward Eggleston, third-generation representative of a pioneer Indiana family, wrote of the frontier with all of the familiarity of one who had been there. His corn shucking bee at Cap'n Lumsden's in the prestatehood era of Indiana comes from his novel, *Circuit Rider* (1874). Another descendant of the frontier, Hamlin Garland, commented that he started writing about "an Iowa corn-husking family, hoping it might please some editor. You see I had the advantage of having spent many days in husking corn." Indeed corn was fundamental to Garland's writings. In his "Among the Corn Rows" from *Main-Travelled Roads* (1891), he tells a well-constructed story of Rob, who returns to Wisconsin from his hundred-acre wheat farm farther west to take a wife. Rob stealthily rescues his childhood acquaintance Julia from her demanding Norwegian parents and her life of corn plowing and other farm drudgery. Their one-day courtship takes place within the shielding embrace of the tall corn.

Garland would later write many stories detailing corn culture in the northwestern prairie and plains country. His *Boy Life on the Prairie* and the four-volume *Middle Border* series (1917–1928) were fiction based on a biography of his family. The novelization of corn reached its zenith in the writings of Hamlin Garland.

Other turn-of-the-century novelists who wove corn into their incidents and descriptive passages included Irving Bacheller and L. Frank Baum. In Bacheller's sentimental *Eben Holden* (1900), Uncle Eb and Willie Brower and dog hide in the tall, dense corn of Vermont and eventually escape their pursuers by a kind of "underground railroad" flight to the valley of the Saint Lawrence. The devoted old Eb is thus enabled to keep custody of the orphaned child, Willie. *The Wizard of Oz* (1900), Frank Baum's tale of fantasy, tells the story of Dorothy and her dog, who are carried by a cyclone from the corn and wheat farms of central Kansas to the land of the Munchkins, who also raise corn and make scarecrows.

Perhaps it was rural nostalgia, a belated awareness that the once-dominant farm life was rapidly falling prey to mechanization and urbanization, a longing to return to the simple gardens of yesteryear. Whatever the causes, rural novels flourished in the teens and the 1920s and 1930s as never before or since in American experience. Often it was a return to the scenes of childhood for the authors, a return to corn as scene and setting. Many of the settings were pioneer; others were modern. Hamlin Garland, portrayer of the northwestern frontier, spanned the decades from 1890 to 1930. By 1900 and for thirty and more years afterward, Mary Johnston and Ellen Glasgow chose a southern setting, rural Virginia, for their flurry of novels. *Lamb in His Bosom* (1933), by Caroline Miller, another novelist of the South, developed many aspects of corn culture, as did Harry H. Kroll's *I Was a Share-Cropper* (1936). The corn lands of prairie and plain were revisited in trilogies by Herbert Quick and Ole Rølvaag (1922–1931), and in *Rim of the Prairie* (1925) and *A Lantern in Her Hand* (1928) by Bess Streeter Aldrich.

The list goes on, as does the subtly implied theme, the sounding of "Taps" for rural American life: Vardis Fisher, Edna Ferber, John T. Frederick, Wellington Rae, Margaret Wilson, Margaret Mitchell, Glenway Wescott, Harold L. Davis, William Faulkner, Willa Cather, Ruth Suckow, Louis Bromfield, Cornelia James Cannon, and Leo L. Ward. John Selby's novel, *Island in the Corn* (1941), treats the northern Midwest, the upper Mississippi region, and the Missouri River Valley. In

The Forest and the Fort (1943), Hervey Allen vividly portrays the awesome timbered expanses and their gradual yielding to axe and plow. "Sex Education," a short story by Dorothy Canfield Fisher, is a subtle psychological analysis of a woman's changing attitudes through varying stages of her maturity after a surprise meeting with a man while she was lost in a dense growth of corn. In Margaret Wilson's *The Able McLaughlins*, Christie mistakes the identity of Squire McLaughlin. Fearing that he is a tramp, she flees through the cornfield. Tom Tryon's surrealistic novel, *The Dark Secret of Harvest Home*, departs radically from the typical fictional treatment of the corn setting.

Fiction writers have dealt with most of the major phases of corn culture in their works—plowing, planting, cultivating, harvesting, grinding, distilling, eating, pipe smoking, but above all, shucking, and concealment among the rows and rows of the towering plant.

Those long, neat rows of waving corn and stately shocks seemed to symbolize and embody the rhythm and meter of poetic lines. Little wonder that the grain, whose very name, *maize*, meant "that which sustains life," would inspire and sustain poetry. Among the earliest of the corn poets was Yale-educated teacher, diplomat, and political writer Joel Barlow. Moved by homesickness while in Paris in 1793, Barlow wrote "The Hasty Pudding." This lengthy descriptive poem had greater spirit and spontaneity than his more formal rhymes. Of his "soul inspiring" hasty pudding, Barlow wrote: "I sing the sweets I know, the charms I feel, / My morning incense and my evening meal." Referring to some "lovely squaw in days of yore," he continued:

> Some tawny Ceres, goddess of her days,
> First learned with stones to crack the well-dried maize,
> Through the rough sieve to shake the golden shower,
> In boiling water stir the yellow flour.
> . . . .
> Could but her sacred name, unknown so long,
> Rise like her happy labors to the son of song,
> To her, to them I'd consecrate my lays,
> And blow her pudding with the breath of praise.

Step by step, page after page, the nostalgic poet traced the corn culture in pastoral lines from plowing to planting to tending to harvesting to the well-known husking bee. Yet Barlow never loses sight of his single goal—to describe in detail the processing of corn to its finest end product, the delectable hasty pudding.

Other poets followed Barlow's lead into the cornfields. Month after month, year after year the farm almanacs fed their poetic lines to readers of the plow and pasture lands of America.

> The thrifty cornfields now display
> The fruit of toil and care;
> Their beaut'ous form and green array,
> The smiles of promise wear.
>
>   (1825)

> The hill tops, crowned with yellow corn,
> Attract the stranger's eye,
> Their blushing fruits the fields adorn,
> And greet the passers-by.
>
>   (1838)

> Hark the rattling husks I hear,
> As Jotham strips the yellow corn;
> Let's kiss thee Pol my pretty dear,
> For't surely cannot be no harm.
>
>   (1800)

While such verses do not qualify as great poetry, they were earthy, grass-roots stuff, close to the heart of rural America. Few poems have been read and enjoyed by such a high percentage of the citizenry.

Henry Wadsworth Longfellow traced the Chippewa Indians through their primitive, spirit-guarded corn culture in his *Song of Hiawatha*, and Bayard Taylor gave poetic recognition to Mondamin, corn god of the Ojibway. But probably no other poet captured the spirit of corn so completely as did Quaker and abolitionist John Greenleaf Whittier in his celebrated stanzas of "The Corn Song." Although his bucolic poetry of the nineteenth-century pastoral scene is now out of favor, his "Barefoot Boy," "Snow Bound," and "The Corn Song" will long echo through the school rooms of farm communities.

> Heap high the farmers' wintry hoard;
> Heap high the golden corn.
> No richer gift hath autumn yet
> Poured out her lavish horn.

Homespun James Whitcomb Riley, sometimes called the Mark Twain of American poetry, captivated middle America with his Hoosier dialect and country philosophy. Among his immortal numbers is "When the Frost is On the Punkin."

> The husky, rusty russel of the tossels of the corn,
> And the raspin' of the tangled leaves,
> As golden as the morn,

> O, it sets my hart a'clickin'
> Like the tickin' of a clock,
> When the frost is on the punkin,
> And the fodder's in the shock.

Constance Woolson's "Corn Fields," Sidney Lanier's "The Waving of the Corn," and Edna Dean Proctor's "Columbia's Emblem" paid further sentimental tribute to the great American grain. Nearly every American, as a youngster, heard the nursery rhyme line, "The sheep's in the meadow, the cow's in the corn," and many residents of the Midwest learned the homely, superstitious couplet of the wart curer: "Barley corn, barley corn, Indian meal, shorts; / Spunk water, spunk water, swallow these warts."

Poets of the twentieth century continued to pay their respects to the country's greatest crop. Ellen P. Allerton wrote about "Walls of Corn" on the once-barren plains of Kansas:

> Who would have dared, with brush or pen,
> As this land is now, to paint it then?
> And how would the wise ones have laughed in scorn,
> Had prophet foretold these walls of corn,
> Whose banners toss on the breeze of morn.

"Corn," by Paul Engle, is a vaulting, abstract poetic treatise of another homesick traveler, who, like Barlow, was far from his native corn country.

> One step, Manhattan to the Alleghenies,
> Myriad lying under mountain laurel,
> One step, the autumn prairie blonde with corn.
> . . . .
> To leave the dare and dazzle of the sun
> Burning the brittle leaves of Kansas corn,
> I wished the blunt and useful speech
> Of men on land where I was born,
> Guessing the pounds of a fatted pig,
> Asking the price of autumn corn.

Carl Sandburg, who scattered his references to corn through *Abraham Lincoln: the Prairie Years*, did not neglect the golden grain in his poetic musings: "Corn Prattlings," "Cornfield Ridge," and "Ripe Corn":

> The wind blows, The corn leans. The corn leaves go rustling. The
> march time and the wind beat is to October drums.

> The stalks of fodder bend all one way,
> The way the last windstorm passed.

Benjamin Wallace Douglas replowed some old poetic ground turned by Whittier in a series of stanzas under the same title, "Corn Song," covering the subject briefly from April furrow to harvest in the frost of autumn.

> The whir of quail, and a ground mouse burrows,
> Under the corn in the long dry furrows:
> There is frost in the air as the golden grain
> Finds its way to the crib again.

Corn and song blended together in more ways than in the titles of verses by Whittier and Douglas. If it was but a step from poetry to music, that step was often the stride of the plowhand plodding and singing through the cornfield behind mules and cultivator, singing for whatever reasons people sing. To soothe the working animals? Farm almanacs advised this. Singing to mitigate the endless toil by giving free vent to the spirit of rhythm and sentiment within as both cowhand and plowhand were wont to do? It was but another step from the chanting and singing of Indians at corn ceremonies to pioneers who bawled such frontier numbers as "Shuckin' of the Corn" at husking bees.

Corn and the farm gave rise to both singing and songs. No songwriter was more prone to turn to the corn theme than Stephen Collins Foster. A northerner by birth and almost lifelong residence, Foster used the southern plantation, slaves, corn, and cotton as the sources of many of his lyrics. Modern developments in race relations, understandably, have caused his songs to pass out of favor. Not all "massas" were as kindly as Foster represented them, and even the kindest would not justify the peculiar institution. Both Foster and his music were products of the times, and they cannot be overlooked by those who would understand the times. "Down in de corn field, hear dat mournful sound" is from his "Massa's in de Cold Ground." From his "Old Kentucky Home" come the words, "The corn top's ripe and the meadow's in the bloom." And poor "Old Ned" "had no teeth for to eat a hoe cake, so he had to let the hoe cake be."

A scattering of corn can be found in other old American songs. "Yankee Doodle," its origin unknown, refers to "the men and boys as thick as hasty puddin'." Sung to the same tune, "Corn Cobs," a ditty of the 1830s read:

> O Aunt Jemima climbed a tree
> And had a stick to boost 'er;
> And thar she sat a throwin' corn
> To our old bob-tailed rooster.
> Corn cobs twist your hair;
> Cart wheels run around you.
> Fiery dragons take you off
> And mortar pestle pound you.

Dan D. Emmett's "Dixie" had its "buck wheat cakes an' Ingen batter" which makes you fat or a little fatter." Romance and corn were brought together by C. Blamphin in "When the corn is waving, Annie dear, our tales of love we'll tell." S. S. Steele struck another southern chord for the omnipresent grain with "On Tom-Big-Bee River, so bright I was born, in a hut made of husks ob de tall yaller corn." At that jumping-off point for the far frontiers, Westport, Missouri, in 1845, people were singing "Lucy Neale," a popular number from a Philadelphia minstrel show:

> Miss Lucy she was taken sick,
> She eat too much corn meal,
> The doctor he did gib her up,
> Alas! Poor Lucy Neale.

Perhaps no more popular song has come out of the South than James Bland's

> Carry me back to Old Virginny,
> There's where the cotton and the corn and taters grow.
> . . . .
> There's where I labored so hard for old Massa
> Day after day in the field of yellow corn.

Godfrey Mark's "Corn Song" has been sung in corn-belt schools for nearly a century. Other examples of corn in song include "That field of corn would never see a plow; / That field of corn would be no good no how" from "Without a Song," and from "Beulah Land," "I've reached the land of corn and wine."

The mill and the still have come in for their share of melodies from "the creek and the old rusty mill, Maggie," to "Down by the old mill stream," and "The creaking old mill on the stream," the last-named a newcomer from the 1930s. From Tennessee came "Moonshine in those Cannon County Hills." "Copper Kettle" is a song of true folk origin:

Get your a copper kettle; get you a copper coil;
Cover with new-made corn mash
And never more you will toil.
You'll just lay there by the juniper,
While the moon is bright;
Watch them jugs a'fillin'
In the pale moon light.

An even better known folk song is "Blue Tail Fly": "Jimmy crack corn an' I don' care; my master's gone away." Husky voices at rural community song gatherings harmonized on a number of popular rounds, including:

Where is John; the old gray hen has left her pen?
Oh where is John; the cows are in the corn again?
Oh J-o-h-n!

Richard Rodgers and Oscar Hammerstein II added a granular dimension to their brand of musical comedy in such lines as "I'm as corny as Kansas in August" from "A Wonderful Guy" in *South Pacific*, and "The corn is as high as an elephant's eye" from *Oklahoma*'s "Oh What a Beautiful Mornin'." *The Tender Land*, Aaron Copland's opera, is set in the midwestern corn and wheat country of the 1930s. Professor Howard Hill, the title character in Meredith Willson's *Music Man*, deplored the hazards of the River City pool hall and dime novels, "hidden in the corn crib." In the musical *Mame* we are told that the lead character could "charm the husk right off of the cob."

Corn as a cultural and artistic form was by no means limited to literary, poetic, and musical modes of expression. Artists and architectural designers, too, reflected in visual representations the importance of this unique product. It is one of the ironies of history that the corn wealth of the American grain belt does much to support museums, art galleries, and other cultural edifices and activities wherein corn is often represented as an art form.

Scores of little-known and unknown illustrators sketched their realistic renditions of corn and its many related tools in numerous farm journals of the early nineteenth century and after. Several papers, Frank Leslie's *Illustrated Newspaper* (1855–1922) and *Harper's Weekly* (1857–1916) created a further demand for illustrators, as did *Century*, *Harper's*, and *Scribner's* magazines. Artists such as Edwin Forbes, A. R. Waud, T. H. Wharton, A. W. Thompson, and Winslow Homer emerged as prominent illustrators through *Harper's Weekly*.

Winslow Homer's THE LAST DAYS OF THE CORN HARVEST, *which appeared in* HARPER'S WEEKLY, *December 6, 1873, portrays the traditional corn and pumpkin harvest.*

The best known of these was Homer, whose illustrations, *Among the Corn* and *The Last Days of the Harvest* were superb in their detail. While working for *Harper's*, Winslow Homer began to paint in oils. He specialized in seascapes and farm scenes, painting such corn-related objects as fodder stacks and barns.

Homer was not alone. The list of rural landscape genre painters who depicted corn subjects and objects is a sizable one. Among Homer's predecessors were William Sidney Mount, noted for his *Boy Hoeing Corn, Fair Exchange No Robbery, Long Island Farmer Husking Corn,* and *Corn*; Jasper F. Cropsey, *Field of Corn Shocks*; and Eastman Johnson, *Corn Husking at Nantucket.* George Inness, Seth Eastman, and lithographers Nathaniel Currier and James Ives also belong in this company of early corn scene painters. Currier and Ives reproduced paintings by such artists as George H. Durrie, whose *Returning to the Farm* depicted the once-popular fodder stacks. In his *Purple Grackle,* John James Audubon applied the same fine touch to the detail of corn husks, grains, silks, and leaves as he did to the depiction of his birds.

The late nineteenth and early twentieth centuries spawned an impressive group of painters who reflected the appeal of Indian corn. Versatile realist George W. Bellows painted *Cornfield and Harvest.* Thomas Hart Benton, a crusty and unconventional artist noted for his three-dimensional effects, produced *Jacques Cartier Discovers the Indians, Roasting Ears, Missouri Landscape* and *Achelous and Hercules.* Olaf Krans's *Woman Planting Corn* depicts twenty-four women in a regimented pattern, each with her hoe and seed bag. Andrew Wyeth showed his genius for painstaking detail and haunting, somber colors in *Corn Seed, Winter Corn,* and his various scenes of old mills.

Iowa's native, Grant Wood, who died of cancer at the peak of his painting career, was the most prolific painter of corn. His works included *Cornshocks, January, Fall Plowing, Stone City, Vegetables, Fertility, Young Corn,* and a huge project consisting of four canvas-covered walls of corn murals in the Corn Room, Hotel Chieftain, Council Bluffs, Iowa. The movie, *Wizard of Oz* depicts a Grant Wood landscape as background for the world of make-believe into which Dorothy travels on the yellow brick road. Kansas-born John Steuart Curry painted many corn scenes, including *Corn, Corn Stalks, Farm Stock, Wisconsin Scene,* and *The Corn Plower.* The rotunda of the Kansas State Capitol at Topeka has a cornstalk decoration by Curry. No listing of corn painters would be complete without mentioning the simple, primitive, but sincere and captivating art of Grandma Moses, who first put her arthritic hands to painting at age seventy-eight. Her *Corn, Checkered House, Little Boy Blue*, and *Pumpkins* are typical of her more than one thousand canvases.

Grain mills had an appeal to painters of rural America like Grandma Moses. George Inness applied his soft touch to several such scenes, including *Mill* and *Mill Pond*. Waud, Forbes, and G. E. Dopler each sketched in this arena. And Regis Gignoux, Jasper Cropsey, Currier and Ives, John L. Barnwell, and Andrew Wyeth were among the many artists who put the big picturesque waterwheels and millhouses to canvas and paper.

The miraculous maize plant is represented in other picture forms. In 1683, just two years after the colony was founded, the Pennsylvania Provincial Council at Philadelphia approved a Kent County seal consisting of three ears of Indian corn. Several corn-belt states have placed corn shocks on their official state seals. Cloth, as well as canvas and paper, has received the imprint of corn art in the colorful decorations on the stage curtain of the Garrick Theater in Chicago.

Like their predecessors the Indians, American pioneers carried corn into the realm of the third dimension, but their sculptured and molded renditions usually had a utilitarian purpose rather than the religious connotation of much Indian art. Corn dolls and corn fodder scarecrows have been discussed in other chapters. Youngsters often fashioned clay figures, using corn kernels for teeth. The corn ear has long been a favorite design for muffin pans, salt and pepper shakers, butter molds, holders for roasting ears, jewelry, and molds for making candles. An ironworker in New Orleans fashioned a wrought-iron fence in a corn-plant design. It still stands around an old rooming house as the Corn House. Corn husks, stalks, cobs, and other corn-related themes are characteristically represented in parade floats all over America.

Among the many designs in the architecture of the government buildings in Washington, D.C., are ears of corn adorning the heads of columns in the Senate Wing of the Capitol Building. This typically American touch has an international flair. It was applied by foreign sculptors under the direction of English architect Benjamin Henry Latrobe, who had been appointed surveyor of public buildings in 1803.

The practical side of dealing with corn was enough to foster a range of architectural designs that changed the profile of America. Granaries, their stark lines slashing prairie horizons, and whirling mills all over the grain country, were geared to both corn and the small grains.

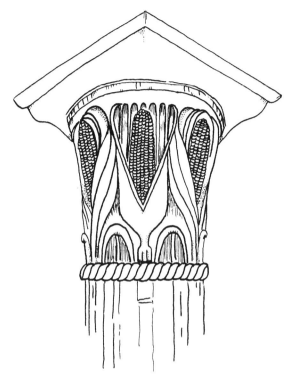

Columns were worked in a design of corn ears at the Capitol, Washington, D.C. They were carved during Jefferson's presidency in 1803.

But Indian corn alone gave America the unique, V-sided cratches or cribs and the tall, statuesque silos. In recent times, barrel-shaped metal tanks have replaced many of the lofty granaries and slotted cribs. These edifices still stand with the silos. However, with rare exception, the old mills have been destroyed or converted to culture centers, such as the museums at Millbach, Pennsylvania, and Wamego, Kansas, and the summer theater house in Jennerstown, Pennsylvania. An old millhouse near Baltimore, Maryland, was used as the headquarters for the George Bush presidential campaign.

Here and there, corn architecture survives purely for art's sake—at the Corn Palace in Mitchell, South Dakota, and at countless dances in the corn country where stalks and ears are arranged in barns and school auditoriums.

To many Indians of the Americas, corn was so necessary to life that it became central to their art, science, oral literature, and religion. These areas of expression seemed blended into one, from the story of

Aztec corn goddess.

the creation or gift of maize to the corn maiden blankets (Navajo), charms, gifts, and medicine men's tricks designed to placate gods and ward off enemies; from the planting ceremony to the green corn dance and ripe corn harvest festival. The deification of corn reached perhaps its most refined forms in sculptured objects of worship, like the beautifully ornate and symmetrical Aztec corn goddess.

American pioneers were also known to chisel graven images, as seen in the stone *Ceres*, goddess of agriculture, on the Old Dutch Mill near Wamego, Kansas, but they did not openly worship stone corn gods. Greater knowledge of science had wiped away some of their ancestors' beliefs in the supernatural. The settlers, as noted earlier, offered their own prayers of hope and thanks to their own unseen gods, however, and often blessed the grain before it was planted and eaten. The difference was only a matter of degree, as they scanned the rough, fibrous pages of their farm almanacs for word about moon phases, days of holiness, and times of feast and celebration.

There were many days to commemorate, most of them brought to America as part of the cultural baggage from the Old World. Lammas Day, August 1, was an import from England. The settlers celebrated it by blessing bread from their first harvest of corn (if they were lucky enough to have an early harvest) or, as was more often the case, blessing the pudding or roasting ears from their green corn harvest. Some farmers observed Rogation Day, three days before Ascension Day, by walking their property lines in worshipful thanks for the precious gift of land. Thanksgiving, Advent, Christmas, Easter, Ascension, and Whitsunday or Pentecost were other festivals related to farm life and farm foods, including that most important of all their crops—Indian corn.

Except for the span of the centuries, a purple gown, and a few embroidered emblems, it might have been an ancient Indian ceremony that February day in 1897. A devout woman in regal garments was dedicating a holy site, a religious monument. Was it so different from an old Mayan or Aztec cult as Katherine Tingley sprinkled grains of corn and drops of wine and oil on the cornerstone of a new building in its first stage of construction? Flags rippled in the breeze, a band played the strains of *Intermezzo*, and the Universal Brotherhood and Theosophical Society of Point Loma, near San Diego, was born.

This cult lasted but half a century—a short life, perhaps, for a religious movement. No doubt the application of corn, wine, and oil was only a dying symbolic ember, a sacrament of communion rather than a show of outright worship of the maize crop. But if American pioneers lacked some of the Indians' reverence for corn, they more than compensated by their worship of its bountiful yields and the material benefits from its ebb and flow in the channels of trade and commerce.

CHAPTER XVIII

Corn on the Hoof: Commerce

While we export fully half of the cotton raised and about one-quarter of our wheat, our exports of corn are but about 2 percent, or one-fiftieth of our annual crop. . . . 80 percent of the crop is consumed in the counties where grown, only 20 percent being shipped outside, and much of this is to adjoining districts for farm use there.

Corn in Kansas, 1929

The foregoing statement might as well have been written in 1829, or for that matter, 1729. The percentages would have differed only slightly before the mid–twentieth century, and not dramatically until the 1970s.

Compared with the cash crops, there was little demand for corn as such in pioneer American commerce. In its raw form, it moved very little. But to the trained observer, maize could be seen in the stream of trade on every hand—in whiskey kegs, pork barrels, milk cans, egg crates, chicken coops, and on the hoof.

Through the centuries, corn was to pioneer commerce what the water table was to thirsty crops: a little-noticed but life-supporting foundation underlying the agrarian economy. It was, to be sure, less flashy than cotton, tobacco, and wheat, the surface streams of farm commerce, with their rapids and falls, their placid pools, sweeping bends, and noisy, colorful movements from place to place. Indian corn was the quiet reservoir.

Why this seemingly subdued commercial role for so dominant a pioneer crop as corn? Why the backstage position relative to the cash crops in the mainstream of trade? There were several reasons. The very success and adaptiveness of corn limited its activity in the forefront of trade. It was the only grain grown in every colony and state. There was no more need to carry corn to colonial New York or to Nebraska than coal to Newcastle or fish to Newfoundland. Yields per acre were so large that corn was only about half as valuable as wheat, bushel for bushel. Thus it was only half as profitable to ship. The large European consumer populations were slow in developing a taste for corn and were much more interested in importing the accustomed wheat throughout the American pioneer era. Corn was uniquely adapted to conversion into other forms—whiskey, pork, poultry, beef, eggs, milk, butter, oil, and starch—and like the coal used for smelting of iron, it lost much of its original weight in the process. A basic rule of economics and industrial location was that the heavy, weight-losing commodities were transported as little as possible in their raw form. Much greater profits could be made from converting corn to whiskey or pork, then shipping these high-value, low-bulk products to market.

These economic laws largely explain why this grain in its raw form played a limited role in distant trade, while corn on the hoof and in the keg were of great commercial value. Cash shortages among the colonists forced them to search for dependable trade items as substitutes for money. Although furs, fish, lumber, and the so-called cash crops were more prominent in trade, corn was soon produced in quantities larger than needed to live. From crash program of survival to cash crop surplus was but a span of a few seasons. Four years after arrival in Massachusetts, enterprising Plymouth colonists traded maize for furs in Maine, and by the early 1630s Jamestown exported the bountiful grain to Caribbean Islands and to some of the New England settlements. John Winthrop's complaint in 1648 that corn was scarce because of shipments to the Azores and the West Indies confirms that the market was expanding. Overseas demands gradually increased, and by the early 1700s shippers were exporting corn regularly to South America, the West Indies, the Atlantic Islands off Iberia and Africa, and to several European countries, particularly Portugal and Spain. In addition, ship captains took on large quantities of maize to feed sailors on the return voyages.

Shipment from farms to docks and ports across the tidewater regions of the Atlantic Coast in colonial times presented few problems. Wagons in New England, boats in the South, and both land and water transportation in the Middle Colonies were adequate, since distances were short and the terrain was not very rugged. It was here on the first frontier that the Powhatan Indians sold corn to Virginia colonists in exchange for trinkets shipped to America in Lord Delaware's ships. Here, too, the Baltimores of Maryland bought one thousand bushels of maize from the Indians and traded it to New England for salt fish and other goods. William Byrd noted that the Nottoway tribe traded corn for alcoholic drinks, which they loved "more than their women and children." Along with their furs and deer skins, the aborigines traded corn to settlers in many frontier areas until the colonists harvested their first season's crops. Beyond the Fall Line into the Piedmont, traders faced increased difficulties, and with widespread post-Revolutionary clearing and settlement of plots across the Appalachians, giant transportation problems emerged. Backwoods producers coped with the obstacles of distance and steep mountain grades in three ways: trading over the high elevations with relatively lightweight, high-value commodities; going with the currents downriver to New Orleans; and eventually using their political muscle and alliances to secure internal improvements or roads, canals, and railroads at public expense.

No one who has puffed up a steep hill needs a detailed explanation why trade across mountains costs more than over the flats or on water. A bushel of corn hauled across the Alleghenies to Philadelphia sold for three or four times the price of the same bushel in Ohio or western Pennsylvania. To and from the western gateway through the first ranges in southeastern Pennsylvania, goods were hauled up and down the Great Valley of the Appalachians in caravans of big, white-sheeted wagons named for Conestoga Creek in Pennsylvania, where they were manufactured. But the roads across the ranges, Kittanning Path, Braddock's Road, Forbes Road, and the Wilderness Road through Cumberland Gap, were not so easily traveled. The caravans that traversed these trails were lines of pack horses. Many of these animals carried their own food—shelled corn, which was fed along the route or stored in secret hideaways for the return trip. Most of the corn that traders transported across the mountains by horseback was in the concentrated form of whiskey, or more accurately, grain alcohol. Each beast could carry two kegs totaling about twenty-four gallons. Some of the horses transported packs of fur.

They followed many pathways, beginning as early as 1784. Typical routes led from the Ohio Valley over Forbes Road to Shippensburg and Philadelphia, along Braddock's Road to Hagerstown and Baltimore, and from Kentucky and Tennessee through the Cumberland Gap to the Great Valley Road or to Richmond Road. On the return trip, pack animals carried small hardware items, bars of iron (bent over the horses' bodies), spices, glassware, beads, tea, chocolate, "looking glasses," and other light goods.

The other trans-Appalachian corn shipments were on the hoof over

Cattle—corn on the hoof—are headed for market in A. R. Waud's sketch for HARPER'S WEEKLY, *October 19, 1867.*

the same general routes, often in the same caravans with whiskey- and fur-laden horses. Stock raisers fattened hogs and cattle on corn in the West, then sold them to professional drovers, such as George and Felix Renick, who took them in great herds over the mountains. Some herds were driven from Ohio to the South where the heavy emphasis on the cash crops like cotton and tobacco caused periodic shortages of staple foods.

Many Americans, when they hear the term *cattle drive*, are con- ditioned to see images of the trails north from Texas—the Chisholm, Western, and Goodnight-Loving corridors of moving, munching beef. Imagine a scene the likes of which surprised Parson Timothy Flint as he headed west across the Alleghenies of Pennsylvania just after the end of the eighteenth century. Topping a rise of the folded range, the traveler saw a cloud of dust drifting through the leaf-clad timber along the trail ahead, looking like smoke from a hundred cannon in frontier battle. There was no ear-splitting boom of guns, but instead the din of barking dogs, shouting drovers on horseback, mooing cattle, and squealing pigs and the dull pound of countless hoofs. A thousand head of corn-fattened hogs and cattle in one herd were moving on the hoof from the Ohio valley frontier to growing populations on the East Coast. The animals would lose some weight on these drives, but since refrigeration was far in the future and salt meat was too bulky to haul across the ranges, it was most economical to let the creatures supply their own power to market.

Downriver to New Orleans the picture was entirely different. A se- ries of great streams, the Ohio, Cumberland, Tennessee, and Mis- sissippi (called the Main Street of the world's grain trade) and their tributaries, provided cheap hauls to the great transshipment port on the Gulf of Mexico. From the New Orleans outlet, shippers traded western products to widely scattered Atlantic ports in both hemi- spheres. Water transportation with the current cost only a fraction of the rates for overland carries across the moutains. Some individual farmers flatboated their own produce, while others engaged in an early form of agricultural cooperative marketing by combining the goods of several families on a single boat. More often than not, these traders were country-store merchants who had taken in farm com- modities on credit or as barter for imperishable groceries such as cof- fee, tea, and sugar, or for hardware, cotton goods, and other manufac- tured items.

Corn-related products, cotton, and tobacco were the chief goods shipped downstream. Scores, even hundreds of barrels might be carried on a single large flatboat, or broadhorn, so called because of its long, hornlike oars used for steering. Such a boat was eight to fifteen feet in width and perhaps forty to a hundred feet long, and of sufficiently shallow draft that it was said to be able to float on a heavy dew. This was one of the original disposable containers. Too cumbersome to be poled, paddled, or pulled by ropes upstream, it would be sold for timber, broken up for a settler's cabin, or abandoned after it had done its ferrying job. Standing at one point along the Ohio River an observer might have counted hundreds of these low, boxy, floating rigs in a single month's time as they silently rounded bends, sped through rapids, and were brushed by overhanging branches. Sometimes a crew would keep its boat moving during the dangerous hours of darkness, or would anchor in midstream rather than risk attacks by river robbers or Indians from the shore. While crew members often escaped in the canoe that was kept on board each boat, the precious cargo, fruit of a year's investment and labor, might be lost. Snags, bars, and rocks also cost time and valuable cargoes.

As for the freight, it consisted largely of corn products. There were rows and rows of whiskey kegs, and barrels and barrels of salt pork or beef from small-scale frontier distillers, butchers, and packers. Bulk corn itself was shipped downriver also (as it still is) but in smaller quantity than cotton, wheat, tobacco, and wheat flour, these having been more in demand for trade by sea to distant ports. Corn was usually shelled and packed in jute bags, barrels, or bins; ground and shipped in barrels; or stowed in the ear on flatboats. Because of the slowness of the boats, moisture and mildew caused much damage to bulk grain shipments. When closed up tightly in damp climates, the grain tended to ooze moisture that became so hot that workers were known to have been burned by the flow when they suddenly opened containers. Some shippers attempted to lessen the humidity by building carefully contained fires on board, but without great success.

Other problems plagued the trade periodically. Before the Louisiana Purchase of 1803 and during the War of 1812, there were occasional interruptions resulting from diplomatic, military, and naval crises. Low water, particularly during the summer months, forced shippers to flood the market by concentrating their consignments during the wetter spring and fall seasons, despite increased mildew and decreased prices. Manufactured goods still came largely from over the mountains because of upriver shipping costs, although keelboats carried limited amounts of cloth, hardware, and other items against the current. After 1811 steamboats helped to solve some of the problems by speeding downriver hauls and chugging upstream with some manufactured goods. Despite the impediments, New Orleans prospered and grew rapidly. As an indication of the overall flow of goods, by the 1830s New Orleans annually received 30,000 to 50,000 barrels of salt pork and some 75,000 barrels of whiskey from the West.

Eastern cities, New York, Philadelphia, Baltimore, Boston, and Charlestown, grew jealous of the expanding trade of thriving New Orleans, and all attempted to capture some of it by improved transportation routes with the frontier. Boston and the two southern cities experimented with roads and rail lines, while Philadelphia made cumbersome connections with the Ohio valley through the Pennsylvania System, an expensive linkage of roads, canals, and inclining tramways. Prior to extensive railroad building, the only highly successful system was New York's Erie Canal, completed in 1825 by the State of New York. It cut the cost of freighting from Buffalo to New York City to one-sixth of the former expense of carrying by wagon and pack animal. Little wonder that the city at the mouth of the Hudson emerged as the nation's largest and wealthiest metropolis. Corn and corn products, along with wheat, were highly significant in this rise to first ranking.

Frontier states, too, attempted to get aboard the bandwagon, or more appropriately, the canal barge, in dealing with their own problems of hauling farm produce and other goods. The canal-digging craze in Ohio, Indiana, and Illinois during the 1830s fell far short of solving their trade problems and left the states' treasuries bankrupt. Gradual extension of the National Road west from Baltimore to Wheeling and across Ohio, Indiana, and Illinois climaxed an era of intensified road building during the quarter century after the War of 1812.

City vied with city; region competed with region; states battled each other for supremacy in the market place; and the three great sections, Northeast, South, and West parlayed, compromised, and horse-traded over such issues as tariffs, internal improvements, bank policies, and public land prices in attempts to make alliances. Henry Clay's American System, an attempt to build a stronger economy nationwide by adopting mutually beneficial economic policies, typified the search for unity in the web of diverse sectional interests. Gradually the growing industrial and labor forces of the Northeast became more closely tied to western farm and fur production, while westerners developed increasing needs for eastern manufactured goods. Nowhere was competition more keen than in the construction of the bewildering network of canals and, later, railroads.

As Clay was scheming to put together his American System for the East, South, and trans-Appalachian West with the aid of natural and artificial arteries of commerce a scattering of tough pioneers was gradually, and often unknowingly piecing together a greater American nation from the western edge of Louisiana Territory to the Rio Grande and the Pacific Ocean. Corn was there, in the commerce of frontier prairie, mountain, desert, and river valley land. Given the corn-related appetites of most overlanders, its role was nearly indispensable.

Following their long-conditioned tastes as well as the advice of various guide books, travelers to Santa Fe, Texas, Oregon, Utah, and California stocked up heavily on cornmeal, parched corn, and bacon, not to mention an occasional jug of corn liquor. Farmers who lived near the chief embarkation points to the Far West, such as Westport and Arrow Rock, Missouri, and Fort Smith, Arkansas, had a better-than-average market for their corn crops, despite the fact that many travelers came supplied from their own farms. Sometimes too well supplied! Robert Brownlee, before leaving Arkansas for the gold mines of California, sold two cribs of corn for twelve cents a bushel, which was only one-fourth to one-sixth the going rate. Overburdened argonauts and other immigrants dumped hundreds of pounds of bacon in the South Pass area of Wyoming in order to lighten the loads of their teams. The Newark overland party brought extra oxen and wagons loaded with corn in the event grass was scarce along the trail to California.

Some of the tribes of Indian Territory during the late 1840s and early 1850s sold corn to travelers for seventy-five cents to a dollar per bushel. Beyond Indian Territory and Westport, however, it was a long way to fresh supplies. In the Southwest from 1846 to 1861, the United States Army spent more money for corn than for any other local product. New Mexico and Utah were sources of some importance to migrants.

Mountain man Jim Bridger taunted Brigham Young with a promise of one thousand dollars for the first ear of corn raised in the Salt Lake Valley, so little confidence had he in the productivity of the region. He underestimated both the Mormons and the soil. Irrigated crops, saved from crickets by the legendary gulls, soon produced bountiful harvests. Argonaut William Kelley wrote of the "magnificent crop" of corn raised by the Mormons in 1849 that was so bountiful, in fact, that passing immigrants purchased large quantities of corn and other crops.

Golden corn helped to widen the gateway to the golden state in other areas, notably New Mexico and the Gila-Colorado region. In the latter location, Robert Brownlee's party from Arkansas traded vermillion to several Indian tribes in exchange for enough corn to proceed to the Mother Lode of California. At Chino in southern California, Isaac Williams ground corn in his mill and sold and traded the meal to passing immigrants. He persuaded many travelers to sign his register, allegedly in the hope of collecting money later through a congressional relief bill.

During the 1850s, Indian traders from Missouri and Arkansas, such as John Plummer, Albert Gallatin Boone, and Charles A. Warfield, engaged in an unusual corn exchange along with their fur, cloth, hardware, and vermillion trade. They bought corn from the tribes, ground it, and sold it back to the Indians. Like the trail breaking and continent crossing that they helped to sustain, these various trans-Mississippi corn trading activities were important out of proportion to the numbers of people, bushels, and dollars involved, although without question they were small compared with the rail transport operations, which, during the same years, were lacing the country's older states and settlements together.

From the beginnings of the nation's first substantial railroad, the Baltimore and Ohio in 1828, iron bands and iron horses steadily overtook, then spectacularly overshadowed canal digging. Railroad mileage caught up with canals by 1840 and tripled canal mileage by 1850. At the outbreak of the Civil War, rails outdid canals nine to one in mileage.

Among western products, the biggest gainers from this octopuslike rail transportation network were corn from the great corn belt of the Central West and wheat from the Northwest and later the Great Plains. Costs of shipment dropped sharply. By 1850 traders shipped freight between Buffalo and New York City for only one-twentieth of the 1817 rate. Time of shipment, too, was sliced to a small fraction of that required by packhorse, wagon, and river and canalboat. At last corn shipments crossed the country in bulk form with a minimum of mildewing.

Railroads did much more than cut down on spoilage and eliminate picturesque packtrains and long drives of hogs and cattle to eastern markets. Mid-continent meat packing expanded by giant strides at the "pork-opolis" of Cincinnati and later at Chicago, Saint Louis, East Saint Louis, and Kansas City. To a much greater extent than before, America's economy became specialized. As the mid–nineteenth century approached, northeasterners could afford to concentrate heavily on industry, relying on western farmers for food supplies. And for their part, many western farmers became highly specialized and commercialized, raising great quantities of corn and wheat, and developing a dependence on eastern manufacturers for many items previously made at home or done without. Between these two sections—Northeast and West—Clay's American System worked.

Thus was born the great American corn belt of the upper Midwest, a child of gleaming metal rails and crosshatched wooden ties, of huge, beadlike strings of freight cars and puffing, smoke-spewing steam locomotives with their clanging bells and mournful whistles in the prairie nights.

There was a new face on the corn country born of this marriage of railway and corn crop, and it was a face that frequently bore a scowl. To their growing dismay, farmers slowly realized that they had largely sacrificed their self-reliance and independence. Those rails that whisked their goods swiftly to distant national and even foreign markets brought back a contagious economic disease. The panacea became a snare; the network an entangling web. Depressions and price fluctuations from faraway lands quickly spread to grain country with an intensity never before experienced. True, there had been early warning signals when England cut American grain purchases with her Corn Laws (more properly, Grain Laws) in 1828 and when the Panic of 1837 struck the West on the heels of its canal-building epidemic. All this seemed like a nightmare past when corn and wheat exports shot upward after British repeal of the Corn Laws in 1846 and distant trade entanglements began to look like genuine sterling. As the century wore on into the seventies and eighties, the tinplate wore off, leaving farm surpluses and the "farm problem" as permanent fixtures. Some railroad lines compounded the difficulties by unfairly juggling freight rates to their own advantage. An important exception occurred during the Civil War. During the forty years prior to that great sectional struggle, the South shifted to a heavier emphasis on cotton. With reduction of the acreage devoted to the main food source, corn, and with the growth of the slave labor force, the lower South gradually swung to a partial dependence on imported corn. Sources disagree sharply as to the extent of this dependence, but it was considerable. Meanwhile the Middle South met most of its own corn needs and the Upper South and West sold surplus corn to the grain-deficient Deep South.

At the onset of the Civil War, intersectional trade was reduced to a trickle. The Illinois Central Line helped to compensate corn farmers for their loss of southern markets by paying them Chicago prices (less transport costs) in exchange for railroad land bought by the farmers. Five hundred additional rail cars and many granaries were built to handle the surge of business. The Illinois Central reshipped the corn by lake steamer to Buffalo and Oswego, New York, and on to eastern and European markets, but lost considerable money on the venture.

With the extension of rails into the corn belt came a conspicuous addition to the landscape: the granary. "Line" elevators for storage were built by large business firms. Local, independent elevator companies competed with the line outfits in the buying of grain from farmers. In time the corn growers saw the railroads, elevator companies, and dealers at the Chicago Board of Trade as middlemen who made off with a hog's share of the profits. Farmers' cooperatives built their own elevators to try to eliminate these middle-marketers, but with limited success. Meanwhile, surpluses continued to mount, and when prices dropped the farmer usually felt the pinch more severely than did industry. Mary Lease of Kansas exhorted farmers to "raise less corn and more hell," and the originally social Patrons of Husbandry, or Granger Movement, took on an increasingly political and economic tint.

Raising less corn was not as simple as it sounded. Corn was corn and wheat was wheat. There was little discrimination of product, and farmers who raised less feared they would be "cutting off their own noses." They were faced with fixed costs in land, livestock, and labor since their animals and family members had to eat whether production was cut or not. Crops were full-year investments and growers had no way of temporarily shutting down the plant to save money. Nor did they have a means of spreading production over the calendar year. Because of the uniform harvest season, all corn matured and hit the market at about the same time, driving prices in a feast-or-famine, free supply-and-demand situation. Unlike manufacturers, who could usually raise or cut production on short notice, farmers, faced with fixed costs and once-a-year seeding time, answered both high and low prices with attempts to increase production. If prices were high, they tried to "up" the output and make a "killing" as a manufacturer would, but when prices dropped, they produced more in order to sustain total income by selling more bushels of grain at the lower rate per bushel. The result was a chronic case of the disease of overproduction regardless of price levels, and price levels tended to continue low. Economist Adam Smith was not a modern grain farmer, else he would doubtless have revised his theories.

Rapid transportation had a twin conspirator in the near banishing of that treasured institution of pioneer America, the single-family subsistence farm. This was mechanization. Technological advances in corn culture emerged at about the same time as modern mechanical improvements in wheat growing, except for corn harvesters, which were not refined until the mid–twentieth century. From an era less than 150 years ago, when nearly 90 percent of the nation's people fed themselves and made their living primarily from the soil, much has changed. Today, 4 percent of the people are farmers, and the number is still dwindling. Yet they feed the other 96 percent and sell such vast quantities abroad that the United States dominates corn and wheat exports more convincingly than the Arab World leads in exports of petroleum. Since Canada sells no corn, the dominance of the United States in corn exports is even greater than in wheat sales, despite Argentina's strong trade position in corn.

In spite of the shrinking percentage of farmers in the population as distraught and displaced sons and daughters of the farm hurried off to urban catch basins, surpluses continue to plague agriculture. Parity prices, subsidy payments for curbing production (the "not-raising-corn" business), and long-term storage capacities and techniques have failed to solve the basic economic cancer that lies concealed beneath those fertile grain fields, overproduction and falling prices. During early 1978, farmers drove their heavy machinery past the president's residence in Washington, D.C., and Plains, Georgia, in protest against the century-old problem, a blight that, but for greatly expanded exports in the last decade, would have been much more serious.

Companion ailments in American society, less obvious perhaps than high yields and tumbling prices, are the overdependence of the many on the few for critically needed food supplies, the skyrocketing prices of farm lands, and perhaps most important, the loss of part of the old-time economic interdependence and unity among the members of each farm family. It is a different world, as surely as if the continent had been tossed aside and replaced by another. A key part of the difference is that corn has at last won the flash and prominence so long denied it in the visible arteries of commerce.

Epilogue

Over a span of more than three centuries, America's farmers watched the weather and the almanac tables, felt the mold of soils to gauge the time for spring sod-turning and planting, and fought the scores of enemies that strove to live at the cost of dying corn. They plowed and hoed the weeds down and viewed with foreboding the drouths and approaching storms of summer. Using simple, almost primitive tools and techniques, they gathered, stored, and fed their life-giving grain. Through all, they labored and laughed, teamed and toiled, and occasionally wept at the turn of misfortune's wheel.

With the explosive burst of modern technology in the postpioneer era, many things about corn and its culture have been rapidly revolutionized. Little more than the plant itself would be recognized by the tillers of centuries gone by. Tractors have replaced horses, mules, and oxen, and the small, artistic hand tools have been supplanted by giant machines that pulverize the soil and tend the crops from seed to stem to storage and grinding. Herbicides, fungicides, and insecticides, those silent, unseen chemical gladiators, do battle with the myriad of tiny enemies and do violence on occasion to the natural environment and the health of humankind.

Two world wars, bracketing a Dust Bowl and Great Depression era, had profound effects on corn. Overcultivation from 1915 to 1919 sowed the seeds of much additional grain, and of the ruinous dust storms of the 1930s. A three-year drouth during the mid-1930s crippled the nation's heartland and put it on an import basis for corn for the only time in the history of the country. Then came World War II. European grain fields became battlefields as they had twenty-five years before, and American grain exports, particularly wheat, corn, and rice, increased.

The Marshall Plan, beginning in 1947, gave corn one of its sharpest gains in the export market. Part of this was accidental, when the Europeans accepted shipments of corn, believing, through continued confusion of terminology, that it was wheat. They were chagrined to receive large shipments of maize, which they still considered fit mainly for livestock. The 1950s and 1960s witnessed huge American grain surpluses, which were only partly absorbed by increased sales to the Netherlands and Great Britain and foreign aid to such developing nations as Egypt and India.

By 1970 United States corn exports had crept up to 12 percent of total production. At about that time the gigantic grain deals with the Soviet Union, Japan, China, and other foreign countries began. By 1980 as the world's meat appetite soared, America's yearly corn production nearly doubled the 1970 figure, while annual export tonnage quadrupled. Now the golden grain is not only big in world food brokerage, but in power brokerage as well.

"Ioway, where the tall corn grows" has become "Ioway, where the corn talk glows"—talk of corn alcohol in gasohol to help replace declining petroleum supplies, talk of corn to redress the country's trade imbalance caused by burgeoning purchases of oil and manufactured products from abroad, talk of corn as a coercive foreign policy weapon, talk of the "great grain robbery," President Carter's January 4, 1980, embargo on the export of seventeen million tons of grain to the Soviet Union as a counter-measure against the invasion of Afghanistan. In early February, 1980, Senator Herman Talmadge of Georgia introduced a bill to offset the domestic effects of this embargo by encouraging more use of corn for making gasohol.

But the talk nationwide is from farms that are much larger and far fewer than in earlier years. As recently as 1950 there were over 5.5 million farms in the United States, averaging 215 acres in size. Now the number is only half the 1950 figure, and the average farm size has increased to 400 acres. Nor will this trend away from small farms be reversible in the foreseeable future. With the current high prices of land and equipment, the start-up costs are far too great for the little farmer. The long era when small farms provided a people's access to self-sufficiency, even when they had limited money to invest, is dead. Industrial labor, service-type jobs, government work, and welfare are the chief alternatives for an increasingly large segment of the population. The single-family corn farm will soon join the brontosaurus, the passenger pigeon, and the ivory-billed woodpecker in that distinguished company of extinct species.

The recent and the present represent a sharp departure from pioneer life and conditions in other ways. In 1978, 30 percent of the nation's farmers received 90 percent of all cash income from farming, yet more than 45 percent of farm subsidies from the government now go to only 5 percent of the farmers, and these the largest in most instances.

The sluggish years have seemingly quickened their pace to match that of a mechanized world, and society has not the time to wring its hands in sorrow over its cultural losses. Yet the vehicle of time cannot begrudge its passengers an occasional nostalgic backward glance. Hamlin Garland, corn farmer and self-styled "son of the middle border," wrote with feeling about the simple life-style and the untarnished prairie lands that had vanished as completely as if they had been but a dream.

> We'll meet them yet, they are not lost forever;
> They lie somewhere, those splendid prairie lands;
> Far in the West, untouched of plow and harrow,
> Unmarked by man's all-desolating hands.

Those warriors of the prairie stand much as of old, topped by blond tassel helmets. Rustling blades bend with winds and blend with distance into walls of green armor. The phalanxes are mightier than ever, and their yield greater. The trees are there, but no longer as dense groves flanking bountiful fields. They are but frail, ruffled fringes, tokens of former days, bordering some of the vast squared carpets of cultivation. Largely because of their fallen ranks, alarming quantities of irreplaceable topsoil blow and wash away each year from the once-virgin grain lands. Unless this continuing erosion can be stopped, American agricultural production faces a drastic decline in the decades ahead.

Gone are the bobbing heads and swishing tails of work animals, and the broad-brimmed hats and bonnets from between the long rows—those fleeting signs of rugged toilers carefully nurturing leafy off-spring. Frost glistens still on golden pumpkins, but no longer among stately shocks. A mechanically rearranged population has forsaken rows of corn and shocks for streets and rows of urban dwellings. Frontiers have been erased, and with them a way of life that was the pioneer family farm.

What was the role of these generations of farm folk in the larger America? They were more than millions of "little people," dabbling in minutiae. Though not all were noble types by any means, in the aggregate they supplied, person for person, a lion's share of the country's moral fiber and family and community stability. If they had not the time or the literacy to pen the lines of philosophy and poetry (or even to read them), if they lacked the time or the schooling to create works of art or the money or inclination to adorn their walls with fine paintings, they found a philosophy in the recurring seasons, a poetry in the symmetry of their tilled rows, an expression of art in the varied natural land and in the functional forms that they sculpted upon it. If they did not often provide the great political heroes or the industrial and trading tycoons who sailed ostentatiously on the current of American history, in a very real sense, they were the mainstream itself. They *were* the larger America.

"We'll meet them yet." For somehow, in the fullness of history's molding processes, in the subconcious of our cultural undergirding, they are with us yet—those phantom shocks in measured rows, the clamorous birds spiraling on set wings to waiting grain fields below, the rhythmic thudding of hominy blocks, the creaking of mill wheels, and the crackling of corncob fires. Perhaps a culture can better cushion its losses if it is aware of what has been lost and if it consciously seeks to replace the losses with enduring values.

Meanwhile, across the face of those vast tilled acres, wandering, tree-lined streams seem to inscribe great question marks for the future of an America devoid of its once-large class of self-reliant small farmers and farm families. Will western historian Frederick Jackson Turner's grave anxieties about the prospects of a nation without the leavening influence of a frontier be borne out? Can the social nourishment of the family farm be found again despite corporate giants and urban "ant farms" of commuting workers? Can an urbanized, leisure-burdened people find substitutes for the corn farm and its rural counterparts that will impart to succeeding generations of children those feelings of self-reliance, of personal worth, of being needed and wanted as hard-working parts of close-knit family and community circles? Whatever the answers to these questions, one observation seems valid: Indian corn and humankind will as closely share the future of America as they have its past, for the life-or-death alliance goes on, whether the implements be gang plow and harrow and combine harvester or planting stick, hoe, and shucking peg.

Bibliography

Books and articles about corn as an agricultural crop and commodity are by no means rare; but references that convey the truer meaning of corn as a way of life appear mainly as fragmentary and obscure comments in thousands of sources on a great variety of subjects. Thus it would be impractical to provide anything approaching a comprehensive listing of sources on the socioeconomic history of corn. Three principal locations of these spottily available comments are almanacs, newspapers, and journals. Since pioneer America was overwhelmingly rural in population and interests, these literary media of the period are slanted toward an agrarian society. Each of the three resources is presented separately, with some representative listings of titles. The corn farmers mentioned in the Preface and the author's twenty years' experience on a rather primitive Ozark corn farm provided additional source material.

Bibliographies

Agricultural History Center, University of California, Davis. *A List of References for the History of Black Americans in Agriculture, 1619–1974.* Comp. by Joel Schor and Cecil Harvey. Washington, D.C.: U.S. Department of Agriculture, 1975.

Bercaw, Louise O., *et al. Corn in the Development of the Civilization of the Americas: A Selected and Annotated Bibliography.* Washington, D.C.: Bureau of Agricultural Economics, 1940. See other works by this bibliographer.

Bowers, Douglas, comp. *A List of References for the History of Agriculture in the United States, 1790–1840.* Davis, Calif.: Agricultural History Center, 1969.

Edwards, Everett E. *A Bibliography of the History of Agriculture in the United States.* Miscellaneous Publication 84. Washington, D.C.: U.S. Department of Agriculture, 1930.

Fusonie, Alan M. *Heritage of American Agriculture: A Bibliography of Pre-1860 Imprints.* Beltsville, Md.: National Agricultural Library, 1975.

Pursell, Carroll J., and Earl M. Rogers. *A Preliminary List of References for the History of Agricultural Science and Technology in the United States.* Davis, Calif.: Agricultural History Center, 1966.

Savage, William W. "Do It Yourself Books for Illinois Immigrants." *Journal of the Illinois State Historical Society,* LVII (Spring, 1964), 30–48.

Schlebecker, John T. *Bibliography of Books and Pamphlets on the History of Agriculture in the United States, 1607–1967.* Santa Barbara, Calif.: ABC Clio, 1969.

Stuntz, Stephen C. *List of Agricultural Periodicals of the United States and*

Canada Published During the Century July 1810 to July 1910. Miscellaneous Publication 398. Washington,. D.C.: Department of Agriculture, 1941.

Taylor, Raymond G. "Some Sources for Mississippi Valley Agricultural History." *Mississippi Valley Historical Review,* VII (September, 1920) 142–45.

Thompson, George F. *Index to Annual Reports of the U.S. Department of Agriculture for the Years 1837 to 1893, Inclusive.* Bulletin 1. Washington, D.C.: U.S. Department of Agriculture, 1896.

Winther, Oscar O. *A Classified Bibliography of the Periodical Literature of the Trans-Mississippi West, 1811–1967.* 2 Vols. Bloomington, Ind.: Indiana University Press, 1961, 1970.

Unpublished Materials

The following manuscript collections are representative of thousands containing rich but scattered references to corn:

Draper, Lyman. Wisconsin State Historical Society, Madison.

Hardeman. Tennessee State Library and Archives, Nashville.

Jackson, Andrew. Library of Congress, Washington, D.C.

Parker, George M. Missouri Historical Society, St. Louis.

Smith, Thomas A. Western Manuscripts Collection, State Historical Society of Missouri, Columbia.

Brownlee, Robert. "Autobiography." Napa County, Calif., October 20, 1892. Courtesy of Patricia Etter, a direct descendant of Robert Brownlee.

Carter, Robert. "Plantation Account and Letter Books of Councillor Carter of Nomini Hall, Westmoreland County, Va., 1759–1792." 16 Vols. Library of Congress, Washington, D.C.

Ruffin, Edmund. Diary, 1856–1865. 14 Vols. Library of Congress, Washington, D.C.

Winter, Joseph C. "Aboriginal Agriculture in the Southwest and Great Basin." Ph.D. Dissertation. University of Utah, Salt Lake City, 1974.

Government Documents

The United States Department of Agriculture and the agriculture departments of the various states have published thousands of items pertaining to corn production and use.

Langworthy, Charles F., and Caroline L. Hunt. *Corn Meal as Food and Ways of Using It.* Washington, D.C.: Government Printing Office, 1914.

———. *Use of Corn, Kafir, and Cowpeas in the Home.* Washington, D.C.: Government Printing Office, 1913.

Thompson, James W. *History of Livestock Raising in the United States, 1607–1860.* Agricultural History Series. Washington, D.C.: U.S. Department of Agriculture, Bureau of Agricultural Economics, 1942.

True, Alfred C. *A History of Agricultural Education in the United States, 1785–1925.* Miscellaneous Publication 36. Washington, D.C.: U.S. Department of Agriculture, 1929.

Texas Agriculture Department. *Corn Culture.* Austin, 1908.

U.S. Department of Agriculture. *Annual Reports.* 1862–1893. Washington, D.C.: Government Printing Office.

———. *Yearbook.* 1894 to the present. Washington, D.C.: Government Printing Office. Formerly listed as *Annual Reports.*

U.S. Department of the Interior. *Agriculture of the United States in 1860.* 8th Census. Washington, D.C.: Government Printing Office, 1864.

U.S. Patent Office, Commissioner of Patents. *Annual Reports.* 1837–1862. Washington, D.C.: Government Printing Office. Beginning in 1849, this publication was divided, Part II dealing with Agriculture. In 1862, it became the *Annual Report* of the U.S. Department of Agriculture.

Vasey, George. *The Agricultural Grasses and Forage Plants.* Washington, D.C.: Government Printing Office, 1889.

Youngblood, B. *Corn Culture for Texas Farmers.* Austin: Texas Department of Agriculture, 1912.

Books and Pamphlets

Abernethy, Thomas Perkins. *The Formative Period in Alabama, 1815–1828.* Montgomery, Ala.: Brown Printing Co., 1922.

Ainsworth, William T. and Ralph M. Ainsworth. *Practical Corn Culture, Written Especially for Corn Belt Farmers.* Mason City, Ill.: W. T. Ainsworth & Sons, 1914.

Allen, R. L. *The American Farm Book.* New York: C. M. Saxton, 1849.

Allen, Richard. *New American Farm Book.* Rev. and enlarged by Lewis F. Allen. New York: Orange Judd, 1882.

The American Farmer's New and Universal Hand-book. Philadelphia: Charles DeSilver, 1856.

Amicus Curiae. *Food for the Million: Maize Against Potato.* London: Longman, Brown, Green, & Longmans, 1847.

Ardrey, P. L. *American Agricultural Implements.* Chicago: The Author, 1894.

Arnold, Lionel K. *Corn As a Raw Material for Ethyl Alcohol.* Ames, Iowa: L. K. Arnold & L. A. Kremer, 1950.

Arnow, Harriet. *Flowering of the Cumberland.* New York: Macmillan, 1963.

Atkeson, Mary M. and Thomas Clark Atkeson. *Pioneering in Agriculture: One Hundred Years of American Farming and Farm Leadership.* New York: Orange Judd, 1939.

Bailey, John M. *The Book of Ensilage; or, the New Dispensation for Farmers.* Billerica: The Author, 1880.

Bailey, Joseph Cannon. *Seaman A. Knapp, Schoolmaster of American Agriculture.* New York: Columbia University Press, 1948.

Bailey, Liberty H., ed. *Cyclopedia of American Agriculture.* 4 Vols. New York: The Macmillan Co., 1907.

Bardolph, Richard. *Agricultural Literature and the Early Illinois Farmer.* Urbana: University of Illinois Press, 1948.

Bartram, William. *Travels Through North and South Carolina.* Philadelphia: James and Johnson, 1791.

Beal, William J. *Grasses of North America.* Lansing, Mich.: Thorp & Godfrey, Printers, 1887.

Beatty, Adam. *Southern Agriculture, Being Essays on the Cultivation of Corn, Hemp, Tobacco, Wheat, etc. and the Best Method of Renovating the Soil.* New York: C. M. Saxton, 1843.

Bennett, Hugh Hammond. *The Soils and Agriculture of the Southern States.* New York: Macmillan, 1921.

Bennett, Richard, and John Elton. *History of Corn Milling.* London: Simpkin, Marshall Co. 1898–1904.

Berguin-Duvallon. *Travels in Louisiana and the Floridas in the Year 1802.* Trans. by John Davis. New York: I. Riley & Co., 1806.

Bidwell, Percy W. and John I. Falconer. *History of Agriculture in the Northern United States, 1620–1860.* New York: Peter Smith, 1941.

Blassingame, John W. *The Slave Community: Plantation Life in the Antebellum South.* New York: Oxford University Press, 1972.

———. ed. *Slave Testimony: Two Centuries of Letters, Speeches, Interviews, and Autobiographies.* Baton Rouge: Louisiana State University Press, 1977.

Bode, Carl, ed. *American Life in the 1840s.* New York: New York University Press, 1967.

Boggess, Arthur C. *The Settlement of Illinois, 1778–1830.* Chicago: Chicago Historical Society, 1908.

Bogue, Allan G. *From Prairie to Farm Belt: Farming on the Illinois and Iowa Prairies in the Nineteenth Century.* Chicago: University of Chicago Press, 1963.

Bond, Beverly W., Jr. *The Civilization of the Old Northwest.* New York: Macmillan, 1934.

Bowman, Melville L. *Corn: Grazing, Judging, Breeding, Feeding, Marketing.* Waterloo, Iowa: The Author, 1915.

Brissot de Warville, Jacques Pierre. *New Travels in the United States of America Performed in 1788.* Trans. from French edition. Dublin, 1792.

Bromfield, Louis. *The Farm.* New York & London: Harper & Bros., 1933.

Brooks, Eugene C. *The Story of Corn and the Westward Migration.* Chicago: Rand McNally, 1916.

Brown, Lester R., with Erik B. Eckholm. *By Bread Alone.* New York: Praeger, 1974.

Brown, Lauren. *Grasses: An Identification Guide.* Boston: Houghton Mifflin 1979.

Brown, Paul. *Twelve Months in New Harmony.* Cincinnati: William Hill Woodward, 1827.

Brown, Samuel R. *The Western Gazetteer; or, Immigrant's Directory.* Auburn, N.Y., 1817.

Browne, Daniel J. *A Memoir on Maize or Indian Corn.* New York: C. M. Saxton, n.d.

Browne, Porter A. *An Essay on Indian Corn.* Philadelphia: J. Thompson, 1837.

Buck, Solon J. *Illinois in 1818.* Springfield, Ill.: Illinois Centennial Commission, 1917.

Buckminster, William. *The Practical Farmer; or, Spirit of the Boston Cultivator.* Boston: Williams, 1840.

Buel, Jesse. *The Farmer's Companion; or, Essays on the Principles and Practices of American Husbandry.* Boston: Marsh, 1839.

Buley, R. Carlyle. *The Old Northwest: Pioneer Period, 1815–1840.* 2 Vols. Bloomington: University of Indiana Press, 1950.

Burkett, Charles W. *History of Ohio Agriculture.* Concord, N.H.: Rumford, 1900.

Burnaby, Andrew. *Travels Through the Middle Settlements in North America, in the Years 1759 and 1760.* London: Payne, 1765.

Burtt-Davy, Joseph. *Maize, Its History, Cultivation, Handling, and Uses.* London, New York: Longmans Green & Co., 1914.

Byrd, William. *Histories of the Dividing Line Betwixt Virginia and North Carolina.* Introduction by William K. Boyd, New Introduction by Percy G. Adams. New York: Dover Publications, 1967.

Caird, James. *Prairie Farming in America.* New York: D. Appleton & Co., 1859.

Campbell, Alfred S. *An Introduction to Country Life.* Princeton: Princeton University Press, 1936.

Carr, Lucien. *The Food of Certain American Indians and Their Method of Preparing It.* Worcester, Mass.: C. Hamilton, 1895.

Carrier, Lyman W. *The Beginnings of Agriculture in America.* New York: McGraw-Hill, 1923.

Chynoweth, James Bennett, and William H. Bruckner. *American Manures; and Farmers' and Planters' Guide.* Philadelphia: Chynoweth, 1871.

Clark, John G. *The Grain Trade in the Old Northwest.* Urbana: University of Illinois Press, 1966.

Cobb, Joseph Buckham. *Mississippi Scenes; or, Sketches of Southern and Western Life.* 2nd ed. Philadelphia: A. Hart, 1851.

Cobbett, William. *A Treatise on Cobbett's Corn.* London: W. Cobbett, 1828.

Collins, Guy N. *The Value of First Generation Hybrids in Corn.* Washington, D.C.: Government Printing Office, 1910. See other volumes on corn by the same author.

Conner, John B. *Indiana Agriculture: Agricultural Resources and Development of the State: The Struggles of Pioneer Life Compared with Present Conditions.* Indianapolis: W. B. Burford, 1893.

Copeland, Robert Morris. *Country Life: A Handbook of Agriculture, Horticulture, and Landscape Gardening.* Boston: Jewett, 1860.

Cotterill, R. S. *History of Pioneer Kentucky.* Cincinnati: Johnson & Hardin, 1917.

Cozzens, Samuel W. *The Marvelous Country; or, Three Years in Arizona and New Mexico.* Boston: Shepard and Gill, 1873.

Crevecoeur, St. John de, Henri L. Bourdin, *et al.*, ed. *Sketches of Eighteenth Century America.* New Haven: Yale University Press, 1925.

Croy, Homer. *Corn Country.* New York: Duell, Sloan, and Pearce, 1947.

Crozier, William, and Peter Henderson. *How the Farm Pays: The Experiences of Forty Years of Successful Farming and Gardening.* New York: Henderson, 1884.

Danhof, Clarence H. *Change in Agriculture: The Northern United States, 1820–1870.* Cambridge, Mass.: Harvard University Press, 1969.

Deane, Samuel. *The New England Farmer.* Worcester, Mass.: I. Thomas, 1790.

Demaree, Albert L. *The American Agricultural Press, 1819–1860.* New York: Columbia University Press, 1941.

Dick, Everett N. *The Dixie Frontier.* New York: Alfred A. Knopf, 1948.

———. *Life in the West Before the Sod House Frontier.* Lincoln, Nebr.: Prairie Press, 1942.

———. *Sod House Frontier, 1854–1890.* New York, London: D. Appleton-Century, 1937.

Dickerman, Charles W. *How to Make the Farm Pay.* Philadelphia: Ziegler, 1870.

Dominguez, Zeferino. *The Modern Cultivation of Corn.* San Antonio: Dominguez Corn Book Publishing, 1914.

Drown, William, and Soloman Drown. *Compendium of Agriculture or the Farmer's Guide, In the Most Essential Parts of Husbandry and Gardening.* Providence: Field and Maxcy, 1824.

Eggleston, Edward. *The Hoosier Schoolmaster.* New York: Orange Judd, 1871.

Emerson, William D. *History and Incidents of Indian Corn and Its Culture.* Cincinnati: Wrightson & Co., 1878.

Emery, Carla. *Old Fashioned Recipe Book: An Encyclopedia of Country Living.* New York: Bantam Books, 1977.

Enfield, Edward. *Indian Corn: Its Value, Culture, and Uses.* New York: D. Appleton and Company, 1866.

Engle, Paul. *Corn: A Book of Poems.* New York: Doubleday Doran & Co., 1939.

Evans, Oliver. *The Young Mill-Wright and Miller's Guide.* Philadelphia: The Author, 1795.

Facklam, Margery. *Corn Husk Crafts.* New York: Sterling, 1973.

Farmer's Almanac and Housekeeper's Receipt Book. Philadelphia: John Simon, 1851.

The Farmer's Centennial History of Ohio, 1803–1903. Springfield: Ohio Department of Agriculture, 1904.

Faux, William. *Memorable Days in America* . . . . London: W. Simpkin & R. Marshall, 1823. Part I reprinted in Reuben G. Thwaites, ed. *Early Western Travels.* 32 Vols. Cleveland: Arthur H. Clark, 1906, Vol. XI.

Finan, John. *Maize in the Great Herbals.* Waltham, Mass.: Chronica Botanica, 1950.

Fletcher, Stevenson W. *Pennsylvania Agriculture and Country Life.* 2 Vols. Harrisburg: Pennsylvania Historical and Museum Commission, 1950–1955.

Flint, James. *Letters from America.* Edinburgh: W. C. Tait, 1822. Reprinted in Thwaites, *Travels,* Vol. IX.

Flint, Timothy. *Condensed Geography and History of the Western States or the Mississippi Valley.* 2 Vols. Cincinnati: E. H. Flint, 1828.

———. *Recollections of the Last Ten Years Passed in Occasional Residences and Journeyings in the Valley of the Mississippi* . . . . Boston: Cummings, Hilliard, & Co., 1826.

Fornari, Harry. *Bread Upon the Waters: A History of United States Grain Exports.* Nashville: Aurora, 1973.

Gardner, Frank D. *Farm Crops: Their Cultivation and Management.* Chicago: John C. Winston, 1918.

Garland, Hamlin. *Boy Life on the Prairie.* New York: London: Harper & Bros., 1899.

———. *Main Travelled Roads.* New York, London: Harper & Bros., 1890.

———. *A Son of the Middle Border.* New York: Grosset & Dunlap, 1917.

Gates, Paul Wallace. *Agriculture and the Civil War.* New York: Alfred A. Knopf, 1965.

———. *The Farmer's Age: Agriculture, 1815–1860.* New York: Holt, Rinehart, and Winston, 1960. See also other works by the same author.

Gaylord, Willis, and Luther Tucker. *American Husbandry.* New York: Harper & Bros., 1854.

Getz, Oscar. *Whiskey: An American Pictorial History.* New York: David McKay, 1978.

Giles, Dorothy. *Singing Valleys: The Story of Corn*. New York: Random House, 1940.

Goffart, Auguste. *The Ensilage of Maize, and Other Fodder Crops*. New York: National Printing, 1879.

Gordon, Asa H. *The Georgia Negro: A History*. Ann Arbor, Mich.: Edwards Bros., 1937.

Gray, Alonzo. *A History of Agriculture*. New York: Crofts, 1925.

———. *Elements of Scientific and Practical Agriculture*. Andover, Mass.: Allen, 1842.

Gray, Lewis C. *History of Agriculture in the Southern United States to 1860*. 2 Vols. Washington, D.C.: Carnegie Institute, 1933.

Greeley, Horace. *What I Know of Farming*. New York: G. W. Carleton & Co., 1871.

Greenberg, David B. *Furrow's End: An Anthology of Great Farm Stories*. New York: Greenberg, 1946.

———. *Land That Our Fathers Plowed*. Norman: University of Oklahoma Press, 1969.

Halstead, Byron D. *Barn Plans and Outbuildings*. New York: Orange Judd, 1904.

Hamy, E. T. *Charles A. LeSeur in North America, 1815–1837*. Kent, Ohio: Kent State University Press, 1968.

Hardeman, Nicholas P. *Wilderness Calling: The Hardeman Family in the American Westward Movement, 1750–1900*. Knoxville: University of Tennessee Press, 1977.

Hariot, Thomas. *Narrative of the First English Plantation in America*. Reprint. London: Bernard Quatrich, 1895.

Harney, George E. *Stables, Outbuildings, and Fences*. New York: Woodward, 1870.

Harris, Gertrude. *Manna: Foods of the Frontier*. San Francisco: 101 Productions, 1972.

Harris, Thaddeus William. *A Treatise on Some of the Insects Injurious to Vegetation*. Boston: State, 1862.

Harris, William Foster. *The Look of the Old West*. New York: Viking Press, 1955.

Hasse, Adelaide R. *Index of Economic Material in Documents of the States of the United States, 1792–1904*. Pub. No. 84, Ky. Washington, D.C.: Carnegie Institute, 1910.

Hatcher, Harlan. *The Buckeye Country: A Pageant of Ohio*. New York: G. P. Putnam's Sons, 1947.

Haworth, Paul Leland. *George Washington, Farmer*. Indiana: Bobbs, 1915.

Hedrick, Ulysses P. *A History of Horticulture in America to 1860*. New York: Oxford University Press, 1950.

———. *History of Agriculture in the State of New York*. Albany, N.Y.: New York State Agricultural Society, 1933.

Henry, William Arnon. *Feeds and Feeding*. Madison: The Author, 1900.

Hilliard, Sam Bowers. *Hog Meat and Hoecake: Food Supply in the Old South, 1840–1860*. Carbondale: Southern Illinois University Press, 1972.

Holbrook, Stewart. *Down on the Farm: A Picture Treasury of Country Life in America in the Good Old Days*. New York: Bonanza Books, 1954.

———. *Machines of Plenty: Pioneering in American Agriculture*. New York: Macmillan, 1955.

Hooper, Edward J. *The Practical Farmer.* Cincinnati: George Conclin, 1840.
Hopkins, Alfred. *Modern Farm Buildings.* New York: McBride, Nast. & Co., 1913.
Howell, Charles, and Allan Keller. *The Mill at Philipsburg Manor, Upper Mills, and a Brief History of Milling.* Tarrytown, N.Y.: Sleepy Hollow Restorations, 1977.
Howells, William G. *Recollections of Life in Ohio from 1813 to 1840.* Cincinnati: Robert Clarke, 1895.
Hunt, Robert L. *Recollections of Farm Life.* San Antonio, Tex.: Naylor, 1965.
Hunt, Thomas F. *The Cereals of America.* New York: Orange Judd, 1904.
Iowa State College and Experiment Station Staff. *A Century of Farming in Iowa, 1846–1946.* Ames: Iowa State College Press, 1946.
Jacobs, Leonard J. *Battle of the Bangboards.* Des Moines, Iowa: Wallace Homestead Co., 1975.
Jefferson, Thomas. *Thomas Jefferson's Farm Book.* Edwin M. Betts, ed. Princeton: Princeton University Press, 1953.
Johnson, Cuthbert W. *The Farmer's Encyclopedia.* Philadelphia: Carey and Hart, 1843.
Johnson, William H. *Principles, Equipment, and Systems for Corn Harvesting.* Westport, Conn.: Avi Publishing Co., 1966.
Johnston, James F. W. *Notes on North America: Agricultural, Economical and Social.* 2 Vols. Boston: W. Blackwood & Sons, 1851.
Jordan, Samuel M. *Corn Growing in Missouri.* Monthly Bulletin, Vol. VII, No. 7, July, 1909. Columbia, Mo.: State Board of Agriculture.
———. *Making Corn Pay.* Springfield, Mass.: Phelps Publishing Co., 1913. See also other works by the same author.
Jugenheimer, Robert W. *Corn Improvement, Seed Production, and Uses.* New York: Wiley, 1976.
Kansas State Board of Agriculture. *Corn in Kansas.* Topeka: B. P. Walker, State Printer, 1929.
Keller, Allan. *Life Along the Hudson.* Tarrytown, N.Y.: Sleepy Hollow Restorations, 1976.
Kelsey, Darwin P., ed. *Farming in the New Nation: Interpreting American Agriculture, 1790–1840.* Washington, D.C.: Agriculture History Society, 1972.
Kendrick, Benjamin B., and A. M. Arnett. *The South Looks at Its Past.* Chapel Hill: University of North Carolina Press, 1935.
Kennedy, Crammond. *Corn in the Blade: Poems and Thoughts in Prose.* New York: Derby & Jackson, 1860.
Latta, William C. *Outline History of Indiana Agriculture.* Lafayette, Ind.: Epsilon Sigma Phi, 1938.
Leaming, John S. *Corn and Its Culture, by a Pioneer Corn Raiser with 60 Years Experience in the Cornfield.* Wilmington, Ohio: Journal Steam Print, 1883.
Leonard, Warren H., and John H. Martin. *Cereal Crops.* New York: Macmillan, 1943.
Liebig, Justus. *Letters on Modern Agriculture.* Ed. by John Blyth. New York: Wiley, 1859.
Lindley, Harlow. *Indiana as Seen by Early Travelers.* Indianapolis: Indiana Historical Commission, 1916.
Lowery, Irving E. *Life on the Old Plantation in Ante-Bellum Days.* Columbia, S.C.: State, 1911.

McCord, Shirley, comp. *Travel Accounts of Indiana, 1679–1961*. Indianapolis: Indiana Historical Bureau, 1970.

McDougle, Ivan E. *Slavery in Kentucky, 1792–1865*. Lancaster, Penn.: Press of the New Era, 1918.

McGinnis R., comp. *The Good Ole Days*, New York: Harper, 1960.

McKeever, William A. *Farm Boys and Girls*. New York: Macmillan, 1912.

McMillen, Wheeler, ed. *Harvest: An Anthology of Farm Writing*. New York: Appleton-Century, 1964.

———. *Land of Plenty: The American Farm Story*. New York: Holt, Rinehart, and Winston, 1961.

———. *Ohio Farm*. Columbus: Ohio State University Press, 1974. See also other works by the same author.

Mallard, Robert Q. *Plantation Life Before Emancipation*. Richmond, Va.; Whittet & Shepperson, 1892.

Mangelsdorf, Paul. *Corn: Its Origin, Evolution, and Development*. Cambridge, Mass.: Harvard University Press, 1974.

Martin, George A. *Farm Appliances*. New York: Orange Judd, 1909.

Maximilian, Prince of Wied. *Travels in the Interior of North America*. London: Ackerman & Co., 1843. Reprint in Thwaites, *Travels*, Vol. XXII.

Mead, Whitman. *Travels in North America*. In three parts. New York: C. S. Van Winkle, 1820.

Melish, John. *Travels in the United States of America in the Years 1806 and 1807, and 1809, 1810 and 1811 . . . .* 2 Vols. Philadelphia: T. & G. Palmer, 1812.

Meyer, Balthasar Henry, *et al. History of Transportation in the United States Before 1860*. Pub. No. 215C. Washington, D.C.: Carnegie Institute, 1917.

Meyer, Roy W. *The Midwestern Farm Novel in the Twentieth Century*. Lincoln: University of Nebraska Press, 1925.

Michaux, François A. *Travels to the West of the Alleghany Mountains*. London: B. Crosby & Co., 1805. Reprinted in Thwaites, *Travels*, Vol. III.

Miller, John C. *First Frontier: Life in Colonial America*. New York: Dell, 1966.

Miracle, Marvin P. *Maize in Tropical America*. Madison: University of Wisconsin Press, 1966.

Montgomery, E. G. *The Corn Crops*. New York: Macmillan, 1915.

Moore, John H. *Agriculture in Ante-Bellum Mississippi*. New York: Bookman, 1958.

Murphy, Charles J. *Indian Corn: A Cheap, Wholesome and Nutritious Food; 150 Ways to Prepare and Cook It*. New York, London: G. P. Putnam's Sons, 1917.

Myers, Robert M. *Children of Pride: A True Story of Georgia and the Civil War*. New Haven: Yale University Press, 1972.

Myrick, Herbert. *The Book of Corn*. New York: Orange Judd, 1904.

Nolan, Aretas W. *Corn Growing: A Manual for Corn Clubs*. Chicago, New York: Row, Peterson, & Co., 1917. See also other works by this author.

North Carolina, Department of Agriculture. *Report on the General Condition of the Agricultural Interests of North Carolina*. Raleigh: Farmer and Mechanic Printers, 1878.

Olmsted, Frederick Law. *A Journey in the Back Country*. New York: Mason Bros., 1860. See also other works by this author.

Parker, Arthur C. *Iroquois Uses of Maize and Other Food Plants*. Albany: University of the State of New York, 1910.

Pearson, Haydn S. *Success on the Small Farm*. New York: McGraw Hill, 1946.

Perl, Lila. *Red-Flannel Hash and Shoo-Fly Pie: American Regional Foods and Festivals.* Cleveland: World, 1965.

Philips, Martin W. *Diary of a Mississippi Planter, January 1, 1840, to April, 1863.* Reprinted in Mississippi Historical Society, *Publications,* X, 305–481.

Phillips, Ulrich B. *Life and Labor in the Old South.* Boston: Little, Brown, 1929.

———. *The Slave Economy of the Old South.* Baton Rouge: Louisiana State University Press, 1968.

Pitkin, Timothy. *A Statistical View of the Commerce of the United States of America.* Hartford: Charles Hosmer, 1816.

Plumb, Charles S. *Indiana Corn Culture.* Chicago: Breeder's Gazette, 1895.

Pool, Raymond J. *Marching With the Grasses.* Lincoln: University of Nebraska Press, 1948.

Pope, John. *A Tour Through the Southern and Western Territories of the United States of North America . . . .* Richmond, Va.: J. Dixon, 1792.

Power, Richard L. *Planting Corn Belt Culture: The Impress of the Upland Southerner and Yankee on the Old Northwest.* Indianapolis: Indiana Historical Society, 1953.

Priest, William. *Travels in the United States of America, Commencing in the Year 1793 and Ending in 1797.* London: J. Johnson, 1802.

Range, Willard A. *Century of Georgia Agriculture, 1850–1950.* Athens: University of Georgia Press, 1954.

Rasmussen, Wayne D. *Readings in the History of American Agriculture.* Urbana: University of Illinois Press, 1964.

Rawick, George P., ed. *The American Slave: A Composite Autobiography.* 19 Vols. Westport, Conn.: Greenwood, 1972.

Redd, George. *A Late Discovery . . . Relative to Fertilizing Poor and Exhausted Lands.* Winchester, Va.: J. A. Lingan, 1809.

Roehl, Louis M. *The Farmer's Shop Book.* Milwaukee: Bruce, 1923.

Rorabaugh, W. J. *The Alcoholic Republic.* New York: Oxford University Press, 1979.

Ruffin, Edmund. *Essays and Notes on Agriculture.* Richmond, Va.: J. W. Randolph, 1855. See also other works by this author.

Sagendorph, Robb. *America and Her Almanacs: Wit, Wisdom and Weather, 1639–1970.* Boston: Little, Brown, 1970. See section on Almanacs.

Salisbury, James H. *History and Chemical Investigation of Maize or Indian Corn.* Albany: C. Van Benthuysen, 1849.

Sanford, Albert H. *The Story of Agriculture in the United States.* Boston: D. C. Heath, 1916.

Sargent, Frederick Leroy. *Corn Plants: Their Uses and Ways of Life.* Boston: Houghton Mifflin, 1899.

Schlebecker, John T. *Agricultural Implements and Machines in the Collection of the National Museum of History and Technology.* Washington, D.C.: Smithsonian Institution, 1972.

———. *Whereby We Thrive: A History of American Farming, 1607–1972.* Ames: Iowa State University Press, 1975.

Schlebecker, John T., and Gale E. Peterson. *Living Historical Farms Handbook.* Washington, D.C.: Smithsonian Institution, 1972.

Schoolcraft, Henry R. *The Indian Tribes of the United States.* 2 Vols. Ed. by Francis S. Drake. Philadelphia: J. B. Lippincott, 1884.

Shaw, Henry Wheeler. *Josh Billings: His Works Complete.* 4 vols. in one. New York: G. W. Carleton, 1883.

————. *Josh Billings' Old Farmers' Allminax*, 1870–1879. New York: Dillingham, 1902.

Shoesmith, Vernon M. *The Study of Corn*. New York: Orange Judd, 1910.

Slight, James, and R. Scott Burn. *The Book of Farm Implements and Machines*. Edinburgh and London: W. Blackwood & Sons, 1858.

Sloane, Eric. *Diary of an Early American Boy*. New York: Ballantine, 1965.

Sprague, George F. *Corn and Corn Improvement*. New York: Academic Press, 1955.

Stampp, Kenneth M. *The Peculiar Institution: Slavery in the Ante-Bellum South*. New York: Knopf, 1963.

Stephens, Henry. *The Book of Farm Buildings: Their Arrangement and Construction*. Edinburgh and London: W. Blackwood & Sons, 1861. See also other works by this author.

Stowell, Marion B. *Almanacs: The Colonial Weekly Bible*. New York: Burt Franklin, 1977. See also section on Almanacs.

Sturtevant, E. Lewis. *Indian Corn*. Albany: Charles Van Benthuysen & Sons, 1880.

————. *Maize: An Attempt at Classification*. Rochester, N.Y.: Democrat and Chronicle Print, 1884.

Thomas, John Jacob. *Farm Implements and Farm Machinery*. New York: Orange Judd, 1883.

Thompson, Dave O., Sr., and William L. Madigan. *One Hundred and Fifty Years of Indiana Agriculture*. Indianapolis: Indiana Sesquicentennial Commission, 1966.

Thwaites, Reuben Gold, ed. *Early Western Travels, 1748–1846*. 32 Vols. Cleveland: Clark: 1904–1907.

Todd, S. Edwards. *The Young Farmer's Manual*. New York: Woodward, 1867.

Treat, Mary L. *Injurious Insects of the Farm and Garden*. New York: Orange Judd, 1882.

Trexler, Harrison A. *Slavery in Missouri, 1804–65*. Studies in Historical and Political Science. Baltimore: Johns Hopkins University Press, 1914.

Tucker, Gilbert M. *American Agricultural Periodicals: An Historical Sketch*. Albany, N.Y.: Privately printed, 1909.

Tunis, Edwin. *Colonial Living*. New York: Thomas Crowell, 1957.

————. *Frontier Living*. New York: Thomas Crowell, 1961.

Vance, Rupert B. *Human Geography of the South*. Chapel Hill: University of North Carolina Press, 1935.

Virginia Company of London. *Records of the Virginia Company of London: The Court Book, 1619–1624*. Susan M. Kingsbury, ed. 2 Vols. Washington, D.C.: Government Printing Office, 1906–1935.

Wahl, Robert. *Indian Corn (or Mais) in the Manufacture of Beer*. Washington, D.C.: Government Printing Office, 1893.

Walden, Howard T., II. *Native Inheritance: The Study of Corn in America*. New York: Harper and Row, 1966.

Wallace, Henry A., and William L. Brown. *Corn and Its Early Fathers*. East Lansing: Michigan State University Press, 1956.

Waltmann, Henry G. *Pioneer Farming in Indiana: Thomas Lincoln's Major Crops*. Washington, D.C.: Smithsonian Institution, 1975.

Waring, George E., Jr. *The Elements of Agriculture: A Book for Young Farmers*. New York: Appleton, 1855.

Washington, George. *The Diaries of George Washington, 1748–1799*. John C. Fitzpatrick, ed. 4 Vols. Boston and New York: Houghton Mifflin, 1925.

Weatherwax, Paul. *Indian Corn in Old America*. New York: Macmillan, 1954.
———. *The Story of the Maize Plant*. Chicago: University of Chicago Press, 1923.
Weaver, Herbert. *Mississippi Farmers, 1850–60*. Nashville, Tenn.: Vanderbilt University Press, 1945.
West, Elliot. *The Saloon on the Rocky Mountain Mining Frontier*. Lincoln: University of Nebraska Press, 1979.
The Western Agriculturalist. Cincinnati: Robinson and Fairbank for the Hamilton County, Ohio, Agricultural Society, 1830.
Weston, George Melville. *The Poor Whites of the South*. Washington, D.C.: Buell and Blanchard, 1856.
———. *The Progress of Slavery in the United States*. Washington, D.C.: Buell and Blanchard, 1857.
Whitaker, James W., ed. *Farming in the Midwest, 1840–1900*. Washington, D.C.: Agricultural History Society, 1974.
Whitten, John H. *The Effects of Kerosene and Other Petroleum Oils on the Viability and Growth of Zea Mais*. Urbana: University of Illinois Press, 1914.
Wickson, Edward J. *The California Vegetables in Garden and Field*. San Francisco: Pacific Rural Press, 1897.
Wilcox, Earley V. *Farmer's Cyclopedia of Agriculture*. New York: Orange Judd, 1905.
Will, George E., and George E. Hyde. *Corn Among Indians of Upper Missouri*. St. Louis: William Harvey Miner, 1917.
Williams, Charles B. *Corn Book for Young Folk*. Boston, New York: Ginn, 1920.
Willoughby, Hugh. *Amid the Alien Corn: An Intrepid Englishman in the Heart of America*. Indianapolis: Bobbs-Merrill, 1958.
Wilson, Everett B. *Early America at Work*. New York: A. S. Barnes, 1966.
Wilson, Harold K. *Grain Crops*. New York: McGraw-Hill, 1948.
Witthoft, John. *Green Corn Ceremonialism in the Eastern Woodlands*. Ann Arbor: University of Michigan Press, 1949.
Woodman, Harold D. *Slavery and the Southern Economy*. New York: Harcourt, Brace, and World, 1966.
Woods, John. *Two Years Residence in the Settlement on the English Prairie in the Illinois Country. . . .* London: Longman, Hurst, Rees, Orme, and Brown, 1822.
Zirkle, Conway. *The Beginnings of Plant Hybridization*. Philadelphia: University of Pennsylvania Press, 1935.

Almanacs

Almanacs are so varied and numerous as to defy a logical pattern of listings. Yet they are invaluable to the researching of agricultural history topics, particularly before the mid–nineteenth century. The researcher should look under the headings Almanac, Farmer's Almanac, and Old Farmer's Almanack. Among the most prominent writers of almanacs (whose names may be found in the Library of Congress and other listings) were Father Abraham, John Armstrong, Andrew Beers, Samuel Burr, Cornplanter, M. Plum, Thomas Sharp, Thomas Spofford, Robert Thomas, Zadock Thompson, Berlin Wright, and David Young. See also the following:

Heartman, Charles F. *Preliminary Checklist of Almanacs Printed in New Jersey Prior to 1850.* Metuchen, N.J.: Privately printed, 1929.

Morrison, Hugh A. *Preliminary Checklist of American Almanacs, 1639–1800.* Washington, D.C.: Government Printing Office, 1907.

The (Old) Farmer's Almanack . . . . New England States. Dublin, N.H.: Yankee, 1793–1831, 1836–1847. Title varies.

The (Old) Farmer's Almanack . . . . Middle Atlantic States. Boston: Little, Brown. Published intermittently in the late eighteenth and early nineteenth centuries.

Wall, Alexander J. *A List of New York Almanacs, 1694–1850.* New York: New York Public Library, 1921.

Newspapers

The following are only a few representative examples among the hundreds of newspapers that directed their messages largely if not primarily to a society of farmers. They range from the late eighteenth century to the mid–nineteenth.

Baltimore, Washington, and Philadelphia. *Niles Weekly Register.*

Boston (Mass.) *New England Courant.*

Charleston *South Carolina State Gazette.*

Cincinnati (Ohio) *Centinel of the North West Territory.*

Franklin *Missouri Intelligencer.*

Lexington *Kentucky Gazette.*

Natchez *Mississippi Gazette*

New Harmony (Ind.) *Free Enquirer.*

New Orleans (La.) *Orleans Gazette.*

New Orleans (La.) *Picayune.*

Pittsburgh (Pa.) *Gazette.*

Rochester (N.Y.) *Moore's Rural New Yorker: An Agricultural and Family Newspaper.*

St. Louis *Missouri Gazette.*

Vincennes (Ind.) *Western Sun.*

Walpole (N.H.) *Farmer's Weekly Museum.*

Washington, D.C. *National Intelligencer.*

Agricultural Journals

The selected listings below cover the period from the early 1800s to the twentieth century, with emphasis on pre–Civil War times. Much additional information is available in the various annual reports, transactions, bulletins, and memoirs of state agricultural agencies and agricultural societies.

Agricultural Museum. Georgetown, D.C.

American Farmer. Baltimore, Md.

Boston Cultivator. Boston, Mass.

California Farmer. San Francisco

Colman's Rural World. St. Louis, Mo.

Country Gentleman. Albany, N.Y.

Culturist. Philadelphia, Pa.

DeBow's Review of the Southern and Western States. New Orleans, La.

Farmer and Gardener. Philadelphia, Pa.
Farmer and Planter. Pendleton and Columbia, S.C.
Farmer's Advocate. Chicago, Ill.
Farmer's Monthly Visitor. Concord, N.H.
Genesee Farmer. Merged with *American Agriculturalist.* Rochester, N.Y.
Indiana Farmer. Indianapolis
Iowa Homestead. Des Moines
Journal of Agriculture. Boston, Mass.
Kentucky Farmer. Frankfurt
Maine Farmer. Winthrop
Massachusetts Plowman. Boston
New England Farmer. Boston
Ohio Farmer. Cleveland
Plow: A Monthly Chronicle of Rural Affairs. New York.
Prairie Farmer. Chicago, Ill.
Rural American. Utica, N.Y.
Rural New Yorker. Rochester, N.Y.
Rural Register. Baltimore, Md.
Southern Cultivator. Athens, Ga.
Southern Planter. Richmond, Va.
Tennessee Farmer. Jonesboro
Western Farmer and Gardener. Cincinnati, Ohio.
Western Rural. Detroit, Mich.
Wisconsin Farmer. Madison
Working Farmer. New York

Articles in Periodicals

Adolf, Leonard A. "Squanto's Role in Pilgrim Diplomacy." *Ethnohistory*, 11 (July, 1964), 247–261.
Anderson, Edgar, and William L. Brown, "The History of Common Maize Varieties of the United States Corn Belt." *Agricultural History*, XXVI (January, 1952), 2–8.
Atherton, Lewis. "The Farm Novel and Agricultural History: A Review." *Agricultural History*, XL (April, 1966), 131–40.
Atkinson, A., and M. L. Wilson, "Corn in Montana." Montana Agricultural Experiment Station *Bulletin* 107, 1915.
Baker, Raymond. "Indian Corn and Its Culture." *Agricultural History*, XLVIII (October, 1974), 94–98.
Bardolph, Richard. "Illinois Agriculture in Transition, 1820–1870." *Journal of the Illinois State Historical Society*, XLI (September, 1948), 244–64, (December, 1948), 415–37.
Bell, Ovid, "Pioneer Life in Calloway County." *Missouri Historical Review*, XXI (January, 1927), 156–65.
Branson, George. "Early Flour Mills of Indiana." *Indiana Magazine of History*, XXII (March, 1926), 20–27.
Britton, Wiley. "Pioneer Life in Southwest Missouri." *Missouri Historical Review*, XVI (October, 1921–January, 1922), 42–85, 263–68, 388–421, 556–79; XVII (1922–23), 62–76, 198–211, 358–75.
Caldwell, Dorothy J. "David Rankin, 'Cattle King' of Missouri." *Missouri Historical Review*, LXVI (July, 1972), 377–94.

Clark, John G. "Antebellum Grain Trade of New Orleans." *Agricultural History*, XXXVIII (July, 1964), 131–42.

Collins, Guy H. "Notes on the Agricultural History of Maize." American Historical Association *Annual Report*, I (1919), 409–29.

Fussel, G. E. "Ploughs and Ploughing Before 1800." *Agricultural History*, XL (July, 1966), 177–86.

Hoyter, Earl W. "Livestock Fencing Conflicts in Early America." *Agricultural History*, XXXVII (January, 1963), 10–20.

Jacka, Jerry. "The Miracle of Hopi Corn," *Arizona Highways*, LIV (January, 1978), 3–15.

Kalm, Pehr (Peter). "Pehr Kalm's Description of Maize: How It Is Planted and Cultivated in North America, Together with the Many Uses of This Crop Plant." Intro. and trans. by Esther Louise Larsen. *Agricultural History*, IX (April, 1935), 98–117.

Kennerer, Donald L. "The Pre–Civil War South's Leading Crop, Corn." *Agricultural History*, XXIII (October, 1949), 236–38.

Klingaman, David C. "The Development of the Coastwise Trade of Virginia in the Later Colonial Period." *Virginia Magazine of History and Biography*, LXXVII (January, 1969), 26–45.

Lindstrom, Diane. "The South's Interregional Grain Supplies." *Agricultural History*, XLIV (January, 1970), 101–14.

Mangelsdorf, Paul C., Elso S. Barghoorn, and Umesh C. Banerjee. "Fossil Pollen and the Origin of Corn." *Botanical Museum Leaflets, Harvard University*, XXVI (September 30, 1978), 237–55.

Marple, Robert P. "Corn-Wheat Ratio in Kansas." *Great Plains Journal*, VIII (Spring, 1929), 79–86.

Mueller, William. "Growing with the Times: The Grim Yield of Modern Farming." *Harper's*, CCLXI (July, 1980), 82–84.

Nelson, Ronald E. "The Bishop Hill Colony and Its Pioneer Economy." *Swedish Pioneer Historical Quarterly*, XVIII (January, 1967), 32–48.

Otto, John Solomon. "A New Look at Slave Life." *Natural History*, LXXXVIII (January, 1979), 8, 16, 20, 24, 30.

Peterson, Walter F. "Rural Life in Antebellum Alabama." *Alabama Review*, XIX (April, 1966), 137–46.

Primack, Martin L. "Farm Fencing in the Nineteenth Century." *Journal of Economic History*, XXIX (June, 1969), 287–91.

———. "Land Clearing Under Nineteenth Century Techniques." *Journal of Economic History*, XXII (December, 1962), 484–97.

Rasmussen, Wayne D. "History of Agriculture in the Northern United States, 1820–1860 Revisited." *Agricultural History*, XLVI (January, 1972), 9–17.

"A Revolution in Iowa: The Corn Picker." *Annals of Iowa*, XL (Winter, 1970), 220–21.

Ross, Earl D. "Lincoln and Agriculture." *Agricultural History*, III (April, 1929), 51–66.

Sabine, David B. "The All-American Grain." *American History Illustrated*, II (January, 1967), 12–17.

Sturdivant, Laura D. S. "One Carbine and a Little Flour and Corn in a Sack: The American Pioneer." *Journal of Mississippi History*, XXXVII (January, 1975), 42–65.

Sturtevant, E. L. "Indian Corn and the Indian." *American Naturalist*, XIX (1885), 225–34.

Vietmeyer, Noel D. "A Wild Relative May Give Corn Perennial Genes." *Smithsonian*, X (December, 1979), 68–76.

Walker, William B. "The Only Man Who Knows Almost Everything About Corn." *Yankee*, XLII (June, 1978), 72–77.

Warntz, William. "An Historical Consideration of the Terms 'Corn' and 'Corn Belt' in the United States." *Agricultural History*, XXXI (January, 1957), 40–45.

Weatherwax, Paul. "The Popping of Corn." *Proceedings of Indiana Academy of Sciences for 1921*. (1922), 149–53.

Wenz, Alfred, "The Heart of the Corn Country." *Dakota Farmer*, XXXVI (1916) 1068–1070

Wheeler, David L. "The Beef Cattle Industry in the Old Northwest: 1803–1860." *Panhandle Plains Historical Review*, XLVII (1974), 28–45.

Withers, Robert Steel. "The Pioneer's First Corn Crop." *Missouri Historical Review*, XLVI (October, 1951), 39–45.

Index

rotation, 30; as Physiocrat, 34; on potato and corn planting, 74; on shelling, 121; as mill owner, 135, 176; on fodder preparation, 152; and repeal of whiskey tax, 158; on storm damage, 205; on corn loss from war, 207; Farm Book of, 216
Jeffries, Charles, 69
Jelly, corncob, 149
Johnny cake, 145
Johnson, Eastman, 227
Johnston, Mary, 27, 219
Jones, Hugh, 51
Journals, Farm, 215–16
Jumping shovel plow, 61

Kalm, Pehr, 25–26, 216
Kansas, 33, 134, 201, 227, 232
Kansas City, Mo., 241
Keelboats, 237–38
Kentucky, 23, 27, 29, 158, 168
Ketcham, Hank, 212
Kieft, William, 158
King Philip's War, 206
Kingsford, Thomas, 179
Knapp, Seaman A., 32, 49
Knives, corn, 53, 105
Kroll, Harry, 219

Labor and labor systems, 35–44
Lampton, William James, 212
Land, corn growing, 4, 20–34
Land Laws, 27–28
Lanier, Sidney, 222
Latter-Day Saints. See Mormons
Leaming, Christopher, 64
Lease, Mary, 242
Lee, General Henry, 156
Legumes. See Beans
Leopold, Aldo, 217
Lincoln, Abraham, 145, 205
Livestock, 150–55, 183, 195–97, 234–36. See also individual species
Locke, John, 33
Locusts, 201
Longfellow, Henry Wadsworth, 70, 221
Lorain, John, 12
Louisiana, 23
Louisiana Purchase, 23, 25, 34
Lowery, Irving, 93
Lye-making, 142–43
Lynde, Major Isaac, 162
Lyon, Mrs. Mark, 3

Maine, 18, 80, 233
Maize thieves. See Blackbirds

Malaria, 26
Malt and malting, 164
Manifest Destiny, 24–25, 27
Mangelsdorf, Paul, 7–8, 217
Marks, Godfrey, 224
Marshall, James W., 133
Maryland, 18, 27, 51, 97, 233
Mash, 155, 164–65
Massachusetts, 17–18, 51, 134, 176, 233
Massasoit, Chief, 17
Maya Indians, 56, 74
Medicine, corn as, 176–78, 181
Meek, Harry H., 168
Metate, 125–26
Mexico, 7, 10, 37, 112, 131
Mexico City, 7
Mice, 107, 192–93
Michigan, 26, 57, 97
Miles, Manley, 97
Miller, Caroline, 219
Mills: animal powered (arrastra), 129–30, 131, 133; hand powered, 126–29; water powered, 128–37; wind powered, 129, 134; tub, 130–31; wet, 134, 136–37
Minnesota, 21, 57, 201
Missions, 133
Mississippi, 61, 74–75
Mississippi River valley, 23–24, 26, 114, 236
Missouri, 69, 74, 161, 170, 213–14
Missouri River valley, 26, 74, 113
Missouri Gazette, 216
Missouri Intelligencer, 31, 216
Missouri Meerschaum Co., 170
Moles, 193
Moline plow, 64
Montgomery, A. K., 74–75
Money, corn as, 175–76
Moon phases. See Planters and planting
Moore, Jerry, 32
Mormons, 24–25, 208, 239–40
Morrill Act, 32, 217
Morris, Francis, 97
Mortar and pestle, 126
Moses, Grandma, 227–28
Mottoes, corn, 20. See also Slang
Mount, William Sidney, 227
Muir, John, 35, 55, 85
Mules, 62, 86–87, 90, 151, 152–53
Mush, cornmeal, 146–47
Music and corn, 223–25

Narraganset Indians, 77, 126, 140
National Intelligencer, 31, 216
Navajo Indians, 37, 74
Nebraska, 20, 73, 102